AF597876

Infection, Immunity, and Genetics

Proceedings of the Ninth Annual Symposium of the Eastern Pennsylvania Branch of the American Society for Microbiology, Philadelphia, 6-7 June 1977

SYMPOSIUM CO-CHAIRMEN: Herman Friedman, Ph.D.
James E. Prier, D.V.M., Ph.D.

SYMPOSIUM COMMITTEE: Toby K. Eisenstein, Ph.D.
T. Juhani Linna, M.D., Ph.D.
Kenneth D. Thompson, Ph.D.

CONTRIBUTORS:

Ciba-Geigy Corporation
Summit, New Jersey

Endo Laboratories, Inc.
Garden City, New York

Lederle Laboratories
Pearl River, New York

Merck, Sharp & Dohme
West Point, Pennsylvania

Microbiological Associates
Bethesda, Maryland

Miles Laboratories, Inc.
Elkhart, Indiana

Smith, Kline & French Laboratories
Philadelphia, Pennsylvania

Previously published volumes in the series of symposia sponsored by the Eastern Pennsylvania Branch of the American Society for Microbiology:

AUSTRALIA ANTIGEN (Proceedings of the Third Annual Symposium, edited by James E. Prier and Herman Friedman)

OPPORTUNISTIC PATHOGENS (Proceedings of the Fourth Annual Symposium, edited by James E. Prier and Herman Friedman)

QUALITY CONTROL IN MICROBIOLOGY (Proceedings of the Fifth Annual Symposium, edited by James E. Prier, Josephine T. Bartola, and Herman Friedman)

MODERN METHODS IN MEDICAL MICROBIOLOGY (Proceedings of the Sixth Annual Symposium, edited by James E. Prier, Josephine T. Bartola, and Herman Friedman)

THE CLINICAL LABORATORY AS AN AID IN CHEMOTHERAPY OF INFECTIOUS DISEASE (Proceedings of the Seventh Annual Symposium, edited by Amedeo Bondi, Josephine T. Bartola, and James E. Prier)

INFECTION CONTROL IN HEALTH CARE FACILITIES (Proceedings of the Eighth Annual Symposium, edited by Kenneth R. Cundy and William Ball)

Infection, Immunity, and Genetics

Edited by

Herman Friedman, Ph.D.

T. Juhani Linna, M.D., Ph.D.

and

James E. Prier, D.V.M., Ph.D.

University Park Press
Baltimore

UNIVERSITY PARK PRESS
International Publishers in Science and Medicine
233 East Redwood Street
Baltimore, Maryland 21202

Typeset by Action Comp Co., Inc.
Manufactured in the United States of America by
Universal Lithographers, Inc.,
and The Optic Bindery Incorporated.

Library of Congress Cataloging in Publication Data

Main entry under title:

Infection, immunity, and genetics.

"Proceedings of the ninth annual symposium of the Eastern Pennsylvania Branch of the American Society for Microbiology, Philadelphia, 6–7 June 1977."
Bibliography: p.
Includes index.
1. Immunogenetics—Congresses 2. Immune response—Congresses. 3. Infection—Congresses. I. Friedman, Herman, 1931- II. Linna, Juhani, 1937- III. Prier, James E. IV. American Society for Microbiology. Eastern Pennsylvania Branch. [DNLM: 1. Immunogenetics—Congresses. QW541 I43 1977]
QR184.I47 596'.02'9 78-16211

ISBN 0-8391-1292-0

Contents

Participants

D. Bernard Amos, M.D.
Department of Microbiology and Immunology
Duke University Medical Center
Durham, North Carolina 27706

D. F. Amsbaugh, B.S.
Laboratory of Microbial Immunity
National Institute of Allergy and Infectious Diseases
National Institutes of Health
Bethesda, Maryland 20014

P. J. Baker, Ph.D.
Laboratory of Microbial Immunity
National Institute of Allergy and Infectious Diseases
National Institutes of Health
Bethesda, Maryland 20014

Anna D. Barker, Ph.D.
Biomedical Sciences Section
Battelle's Columbus Division
Columbus, Ohio 43201

Joseph T. Blake
Laboratory of Clinical Investigation
National Institute of Allergy and Infectious Diseases
National Institutes of Health
Bethesda, Maryland 20014

Walter S. Ceglowski, Ph.D.
Department of Microbiology
Pennsylvania State University
University Park, Pennsylvania 16802

Anthony J. Dennis, Ph.D.
Biomedical Sciences Section
Battelle's Columbus Division
Columbus, Ohio 43201

Peter C. Doherty, Ph.D.
The Wistar Institute of Anatomy and Biology
36th Street at Spruce
Philadelphia, Pennsylvania 19104

Toby K. Eisenstein, Ph.D.
Department of Microbiology and Immunology
Temple University School of Medicine
Philadelphia, Pennsylvania 19140

Herman Friedman, Ph.D.
Department of Microbiology and Immunology
Albert Einstein Medical Center and Temple University School of Medicine
Philadelphia, Pennsylvania 19140

Ann E. Gabrielsen, Ph.D.
New York State Kidney Disease Institute
New York State Department of Health
Division of Laboratories and Research
Albany, New York 12201

Kathleen Kelly
Department of Medical Microbiology
College of Medicine
University of California at Irvine
Irvine, California 92717

Joseph H. Kite, Jr., Ph.D.
Department of Microbiology
State University of New York at Buffalo
Buffalo, New York 14214

Michael Largen, Ph.D.
Department of Medical Microbiology
College of Medicine
University of California at Irvine
Irvine, California 92717

T. Juhani Linna, M.D., Ph.D.
Department of Microbiology and Immunology
Temple University School of Medicine
Philadelphia, Pennsylvania 19140

Paul H. Maurer, Ph.D.
Department of Biochemistry
Jefferson Medical College
Thomas Jefferson University
Philadelphia, Pennsylvania 19107

B. Prescott, Ph.D.
Laboratory of Microbial Immunity
National Institute of Allergy and Infectious Diseases
National Institutes of Health
Bethesda, Maryland 20014

James E. Prier, D.V.M., Ph.D.
Director, Centre Square Veterinary Clinics
Centre Square, Pennsylvania 19422
Co-Director, Penndel Medical Laboratories
Ardmore, Pennsylvania 19003
Visiting Professor of Microbiology
Harcum Junior College
Bryn Mawr, Pennsylvania 19010
Director of Comparative Medicine Laboratory
Philadelphia College of Osteopathic Medicine
Philadelphia, Pennsylvania 19131

John M. Rice, Ph.D.
Biomedical Sciences Section
Battelle's Columbus Division
Columbus, Ohio 43201

Alan S. Rosenthal, M.D.
Laboratory of Clinical Investigation
National Institute of Allergy and Infectious Diseases
National Institutes of Health
Bethesda, Maryland 20014

Lanny J. Rosenwasser, M.D.
Laboratory of Clinical Investigation
National Institute of Allergy and Infectious Diseases
National Institutes of Health
Bethesda, Maryland 20014

J. A. Rudbach, Ph.D.
Department of Microbiology
University of Montana
Missoula, Montana 59801

Donald H. Silberberg, M.D.
Department of Neurology
University of Pennsylvania School of Medicine
Philadelphia, Pennsylvania 19104

Steven C. Specter, Ph.D.
Department of Microbiology
Albert Einstein Medical Center
York and Tabor Roads
Philadelphia, Pennsylvania 19141

P. W. Stashak
Laboratory of Microbial Immunity
National Institute of Allergy and Infectious Diseases
National Institutes of Health
Bethesda, Maryland 20014

Stephen I. Vas, M.D., Ph.D.
Department of Microbiology and Immunology
McGill University
Montreal, Quebec, Canada

James Watson, Ph.D.
Department of Medical Microbiology
College of Medicine
University of California at Irvine
Irvine, California 92717

R. M. Williams, M.D., Ph.D.
Department of Medicine
Peter Brent Brigham Hospital
Sidney Farber Cancer Institute
Harvard Medical School
Boston, Massachusetts 02115

Edmond J. Yunis, M.D.
Division of Immunogenetics
Sidney Farber Cancer Institute
Harvard Medical School
Boston, Massachusetts 02115

Preface

Higher organisms have evolved in concert with microorganisms. An understanding of both the symbiotic relationship of animals with bacteria as well as the parasitic relationship between various microorganisms and the multicellular organisms they are colonizing or infecting has developed during the last century. A functioning immune response system is a necessary prerequisite for survival of higher animals such as mammals, birds, reptiles, and fishes in an environment consisting of an overwhelming variety and number of microorganisms, many of which may be specific pathogens. The efficacy of the immune system in preventing infectious disease is illustrated by the ability of healthy individuals with normal immune responses to ward off infections as compared to those with defects in various components of humoral or cellular immune competence.

It seems remarkable to note that until about two or three decades ago very little was known about molecular or cellular mechanisms involved in the immune response. An understanding of the nature and structure of immunoglobulins was not developed until the 1950s. It was widely recognized as early as the beginning of this century that cellular immunity is important in protection from infectious diseases. However, it was not until the last decade or two that a fuller understanding of the important role of cell-mediated immune responses working in concert with the humoral, antibody forming arm of the immune system, in protection from infectious microorganisms, has developed.

Another "revolution" concerning basic understanding of the immune system occurred when the genetic control of immune responsiveness and its relationship to the major histocompatibility system because evident. It seems clear that it is important to understand not only antibody formation and cellular mechanisms of immunity during infectious diseases but the genetic basis of immunity, if a full knowledge of the host-parasite relationship is to develop. An analysis of the histocompatibility system also is important in the areas of infectious disease, mechanisms of resistance to tumors, and certain immunopathological diseases. For example, it has been recently demonstrated in both animals and man that certain virus infections may be more common in individuals with specific histocompatibility antigens. Furthermore, genetically determined immune responsiveness to bacterial antigens also has been examined in recent years. It has been observed in mice, rabbits, and guinea pigs that antibody formation and cellular immunity to a wide variety of microbial antigens may be related to genetic factors. The apparent close functional association between histocompatibility antigens and viral and other neoantigens for the recognition of "non-self" is particularly fascinating.

It was the purpose of the symposium upon which this book is based to focus attention on the genetic control of the immune responses to microbial agents, both in animals and man. It was anticipated that providing a forum for presentation of newer knowledge and data concerning infections, immunity, and genetics would encourage additional cross-fertilization of ideas and concepts. Those who investigate the mechanisms of resistance to infectious diseases per se can benefit greatly by understanding some of the basic principles of immunology, especially the newer findings concerning mechanisms involved in both the efferent and afferent limbs of immunity. Furthermore, those who are involved in genetic aspects of immunity and infectious diseases certainly gain by understanding the complexity of microbial antigens, as well as of host-parasite relationships in general. It is the expectation of the organizers and participants of the symposium that the material presented will provide a valuable starting point for further studies and investigations in the exciting area of microbial infections and immunity.

Herman Friedman
T. Juhani Linna
James E. Prier

Infection, Immunity, and Genetics

GENETIC CONTROL OF IMMUNE RESPONSES: INTRODUCTION AND BACKGROUND

Paul Maurer

In the past few years many crucial discoveries have placed the major histocompatibility complex and gene products of the major histocompatibility complex (MHC) on a much broader and higher level of biologic importance than was heretofore considered. Several factors have been responsible for advancing our knowledge and understanding of the major histocompatibility complex and for providing increasingly sophisticated techniques which are now making new discoveries possible.

The first major advance was associated with the production, availability, and genetic characterization of a large number of inbred strains and congenic strains of mice, which differed only within the major histocompatibility complex or H-2 gene complex, and of H-2 recombinant mice. These developments were largely associated with the work of Gorer and Snell, and in which the first contributor, Dr. Bernard Amos, also played a role. The studies of the H-2 complex in mice laid the groundwork and provided further insights into comparisons with the MHC of other species, particularly of the HLA complex of man.

A significant development which depended on the above mice was that the genetic control of the immune responses of inbred and congenic mice against proteins, synthetic polypeptides of amino acids, and viruses could be shown to be associated or linked to the H-2 region, and more particularly in recent years to a number of areas within the H-2 region termed the I region.

The second major advance occurred when techniques were developed in which cells of the immune system could be isolated and identified by certain surface markers. These techniques then led to the development of a number of assay procedures for studying the nature of these lymphocytes involved in the immune response. It was soon realized that indeed there was an important kind of collaboration associated with the interactions between T lymphocytes, B lymphocytes, and the appropriate macrophages. It is in this latter area that Dr. Allan Rosenthal has made a number of specific contributions.

The third series of advances was associated with the development of antisera for the gene products of the defined loci of the major histocompatibility complex. These antisera allowed further identification on lymphocytes and macrophages of diverse surface macromolecules which were concerned with the mechanisms of recognition and regulation, as indicated above.

The H-2 and HLA regions have now grown in importance in the past years. The genes controlling responses not only against natural and synthetic antigens but the susceptibility to viral, bacterial, and autoimmune disease, which we will read about later in this volume, have been mapped and shown to be associated with these loci.

Infection, Immunity, and Genetics
Edited by Herman Friedman, T. Juhani Linna, and James E. Prier

HLA AND DISEASE: A SURVEY OF THE HUMAN HISTOCOMPATIBILITY COMPLEX FOR MICROBIOLOGISTS

D. Bernard Amos

The HLA system is now becoming the center of attention of physicians and microbiologists because it is associated with many diseases, including some infectious diseases. It continues to interest surgeons because of its relevance to transplantation. This may be because of direct involvement of products of genetic loci within the HLA system in disease susceptibility or graft rejection, or because such loci are close to HLA and HLA antigens therefore serve as convenient markers.

HLA is unlike any other known genetic system in man. It is the major human histocompatibility complex (MHC) and as such is very similar to the MHC of other animals and birds. Because of this similarity it must have been advantageous, in an evolutionary sense, to conserve the constituent genes on a short segment of chromosome, and powerful gene interactions must depend upon this conservation. The assemblage of genes on the sixth chromosome comprising the MHC is referred to as the HLA haplotype. The HLA haplotype includes at least three loci that code for integral membrane molecules (Figure 1). These molecules, HLA-A, HLA-B, and HLA-C, are found on the surface of most somatic cells and are especially prominent on cells of the reticuloendothelial system. The A and B locus products are glycoproteins with a mass of 45,000 daltons and are found on the cell surface associated with a smaller peptide of 12,000 daltons which, when free in the serum or urine, is known as β_2 microglobulin

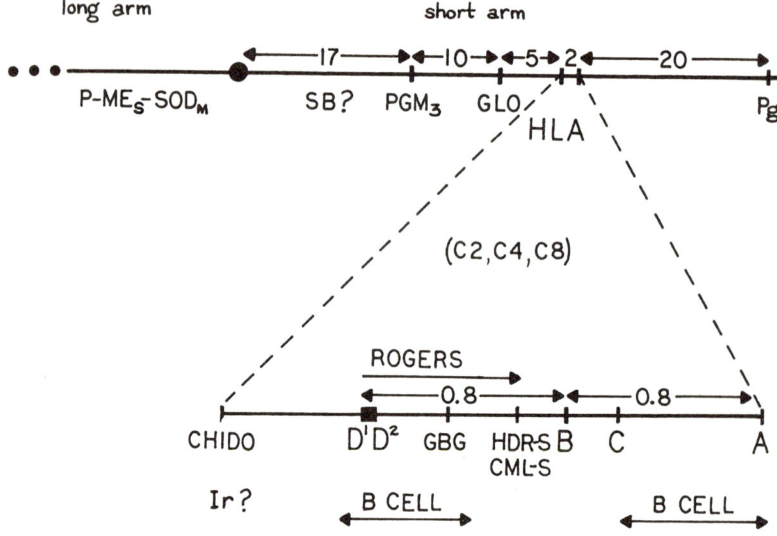

Figure 1. Chromosome 6. The HLA haplotype region.

(β_2M). The β_2M molecule, which is noncovalently bound to the HLA molecule, seems to stabilize the conformation of the HLA site. No variability has yet been found in β_2M from different individuals and it is probable that neither the β_2M nor the carbohydrate plays any direct role in the antigenic specificity of HLA.

Although none of the HLA-A, B, or C molecules is resistant to extreme changes in physical conditions, HLA-C seems to be the least stable and no clear description of its chemistry is available. In contrast, some A and B series antigens have been isolated, characterized, and partially sequenced both from the N terminal end and internally (1,2). The HLA-A locus codes for at least 20 allelic variants. HLA-B also codes for at least 20 alleles, and although only five alleles at HLA-C have been formally recognized, other variants are known (Table 1). Besides being unstable in vitro, HLA-C may be unstable in vivo because few reliable anti-HLA-C antisera are available. As each variant is described it is assigned a number (e.g., HLA-A2, HLA-B7). A "w" preceding a numerical designation signifies that it is a provisional antigen still under study (e.g., HLA-Aw31). It is probable that one or more HLA-like molecules remain to be described as certain antisera give unexplained reactions (D. B. Amos and M. Woodbury, unpublished material).

HLA-A antigens are usually detected on peripheral blood

Table 1. Complete listing of recognized HLA specificities and previous symbols[a]

New	Previous	New	Previous	New	Previous	New	Previous
HLA-A1	HL-A1	HLA-B5	HL-A5	HLA-Cw1	T1	HLA-Dw1	LD 101
HLA-A2	HL-A2	HLA-B7	HL-A7	HLA-Cw2	T2	HLA-Dw2	LD 102
HLA-A3	HL-A3	HLA-B8	HL-A8	HLA-Cw3	T3	HLA-Dw3	LD 103
HLA-A9	HL-A9	HLA-B12	HL-A12	HLA-Cw4	T4	HLA-Dw4	LD 104
HLA-A10	HL-A10	HLA-B13	HL-A13	HLA-Cw5	T5	HLA-Dw5	LD 105
HLA-A11	HL-A11	HLA-B14	W14			HLA-Dw6	LD 106
HLA-A28	W28	HLA-B18	W18				
HLA-A29	W29	HLA-B27	W27				
HLA-Aw19	Li[b]	HLA-Bw15	W15				
HLA-Aw23	W23	HLA-Bw16	W16				
HLA-Aw24	W24	HLA-Bw17	W17				
HLA-W25	W25	HLA-Bw21	W21				
HLA-Aw26	W26	HLA-Bw22	W22				
HLA-Aw30	W30	HLA-Bw35	W5				
HLA-Aw31	W31	HLA-Bw37	TY				
HLA-Aw32	W32	HLA-Bw38	W16.1				
HLA-Aw33	W19.6	HLA-Bw39	W16.2				
HLA-Aw34	Malay 2	HLA-Bw40	W10				
HLA-Aw36	Mo*	HLA-Bw41	Sabell				
HLA-Aw43	BK	HLA-Bw42	MWA				

Reprinted from the WHO-IUIS Terminology Committee, Nomenclature for Factors of the HLA System, by permission.

[a]The previously reserved specificities W4 (4a) and W6 (4b) remain w4 and w6. These specificities are closely associated with the B locus.

[b]HLA-Aw19 includes at least HLA-A29, Aw30, Aw31, Aw32, Aw33, and Aw34(?).

lymphocytes using a cytotoxicity test of which there are three commonly used procedures. Simplest is the Kissmeyer-Nielsen method (3) in which lymphocytes, antiserum, and complement are mixed together on a special slide under oil. The Terasaki method, also known as the NIH method, is a two-stage procedure in which the lymphocytes are sensitized with antiserum in a special microtiter plate and complement is added later (4). The Amos method or Modified NIH technique is a three-stage procedure in which the cells are washed after sensitization and before the addition of complement (5). In all three methods the presence of the antigen is inferred through damage to the cell membrane, as indicated by the uptake of a vital dye. The three procedures have different degrees of sensitivity.

Most antisera used in HLA testing are obtained by screening the serum of women post partum, soon after delivery or later as blood donors. Up to 30% of parous women have a detectable anti-HLA antibody; the incidence drops off soon after delivery for many but not all multipara, and antibody is found in only about 10% of blood bank donors. Of the antibodies found, less than 10% are of minimum acceptable strength and specificity. Screening for antisera is thus a tedious and costly procedure. A very valuable second source of antibody, especially for some of the very unusual specificities, is therefore the volunteer deliberately immunized with lymphocytes from a selected donor. In some instances antibodies are restimulated in mothers past child-bearing age by sensitizing with cells from a child. Some antisera are also available commercially. These sera are often blends of different bleedings of the same or of a different donor and are of restricted value for research, and sera to detect "difficult" antigens are not commercially available. Most laboratories have ample reserves of a limited number of lyophilized or frozen antisera which they augment either from a serum bank maintained by the National Institutes of Allergy and Infectious Disease or by donations from other laboratories. Fortunately the sera are generally very stable and may retain activity for months at room temperature if clarified by homogenization with 30% v/v Freon 113, centrifuged, and ultrafiltered.

Although cross-reactivity (as explained below) may occur, HLA typing is relatively simple for the specificities designated HLA, e.g., HLA-A2, in Table 1. It is, however, usually essential to include at least two sera for each antigen because occasional false negative reactions are encountered. That they are false negative reactions can be checked for by absorption with the suspected cell and retesting on a known positive cell. Typing for specificities designated "w," e.g., HLA-Aw31, in Table 1 is more difficult for two reasons. Available sera are less reliable and cross-reactivity is often a serious problem. Although this is not an appropriate place for a detailed discussion, a few illustrations of cross-reactivity can be given. HLA-B7 and HLA-B27 are highly cross-reactive; an antibody detecting HLA-B7 will often also react with cells carrying B27, and vice versa. The cross-reactivity, e.g., anti-HLA-B27, is usually weaker than the specific reactivity, anti-B7, except when the cell being typed comes from a donor homozygous for B27. For this reason a complete phenotype rather than the identification of a single specificity is highly desirable.

There are families of cross-reactivity; almost all are between alleles of the same locus. (Other examples include: at the A locus, A2 and A28, A3 and A11, A10 and Aw25 or Aw26; at the B locus, B5 and Bw35, B8 and B14, B13 and Bw40, etc.) One special form of cross-reactivity involved two unusual alleles, Bw4 and Bw6, also frequently known as *4a* and *4b*, respectively. These seem to form a two-allele system at the B locus (6). Both are high frequency alleles. Anti-Bw4 sera react with cells from all (or almost all) donors carrying antigens B5, B12, B13, BHR, Bw15, Bw38, and Bw17. Anti-Bw6 sera react with B7, B8, Bw41, Bw15, Bw39, B18, etc. Interestingly, anti-Bw4 reacts with cells from some Bw16 and some Bw22 whereas anti-Bw6 reacts with the remaining donors in each group. Most of the more specific antisera to Bw16 or Bw22 cannot make this discrimination but other sera, sometimes cross-reacting, can, so the distinction or "split" is meaningful. Many of the other specificities such as B5 can also be split by selected antisera and the list of specificities given in Table 1 is therefore incomplete. Immense complexity is also revealed when populations other than Caucasians are tested using Caucasian sera (7). Many pairs of sera giving highly concordant reactions in Caucasians show great divergence when tested on Blacks or Orientals (8). Some specificities are also found almost exclusively in Blacks or Orientals.

The enormous amount of detailed information about HLA serology might be considered meaningless except to geneticists and chemists if it were not for the transplantation and disease association aspects of HLA. For example, is a donor with the subtype of Bw16 classified as Bw38 by virtue of reactivity with anti-Bw4 truly compatible with a donor classified as Bw39 and reactive with anti-Bw6? This is at present an unanswered question. The correct identification of specificities may also have relevance to disease associations. This is most clearly indicated by reference to B27 and its association with the rheumatoid disease ankylosing spondylitis (AS). AS is an inflammatory disease of the spine predominantly affecting young males. If untreated, great deformity results. Many studies in Japanese as well as in Caucasians in Europe and in the United States have shown that approximately 90% of the affected subjects type as B27 positive. Family studies suggest that the antigen and susceptibility to AS segregate together. The association between A3 and the iron storage abnormality hemochromatosis is nearly as strong. As mentioned earlier, B27 and B7, A3 and A11 are very strongly cross-reactive and it might be thought that B27 and B7, A3 and A11 merely reflect very minor variants of rather academic interest. However, B7 shows no significant association with AS and A11 is not elevated in hemochromatosis, so the antigenic differences reflect a biologic and functional difference. It is therefore believed that the HLA region includes, besides the A, B, and C loci, genes relevant to disease. It is known that the murine MHC analogue, H-2, includes genes regulating immune responses and also susceptibility to tumor induction by several viruses. One gene, for ragweed sensitivity (RWS), has been assigned to C6, the chromosome that carries the structural gene for HLA. The obvious next question is then: why is AS associated with B27 and not with any other specificity, such as B7? The answer is not yet known, but it is probably a reflection of a very complex phenomenon called linkage disequilibrium.

Refer to the haplotype (Figure 1). Crossing over is fairly frequent between the HLA-A and HLA-B and between HLA-B and HLA-D. We therefore believe that there is an appreciable distance between the loci, and that in time combinations of alleles at the three loci would become random. Each haplotype expresses two antigens, an A locus antigen and a B locus antigen. If two alleles of B, for example *B7* and *B12*, had approximately the same frequency in the population (which they do), they should be found at the same frequency with any specified A locus allele. This is not true. *A3-B7* is more than ten times as common as *A3-B12;* conversely, *A2-B12* is twice as frequent as *A2-B7.* The phenomenon is known as linkage disequilibrium, and it is observed not only between A and B but also between B and D and especially between B and C. Examples are *B7* and *Dw2, B8* and *Dw3,* and the disequilibrium is exceptionally strong between *Bw35* and *Cw4.* It is believed that linkage disequilibrium will, in time, prove of great predictive value for transplantation and for disease. We do not know for sure that incompatibility at the A locus or B locus antigens is in itself responsible for graft rejection. There are strong reasons for supposing that the real histocompatibility locus lies between HLA-B and HLA-D (9,10). Similarly, although *HLA-B8* and *Dw3* are both highly associated with the disease gluten sensitive enteropathy, the gene involved in disease susceptibility is probably neither of these but may be coded for in the haplotype region near HLA-D. The better defined the antigenic markers and the greater the degree of disequilibrium between the marker and the susceptibility gene or histocompatibility locus, the greater the predictive value of HLA typing.

At present the HLA markers are mainly of value to the clinician in the differential diagnosis of rheumatoid diseases, including the post infection arthropathies (AS), anterior uveitis, and allied diseases. Within the family the markers have predictive value for transplantation and diseases such as hemochromatosis (one of the few diseases associated with an A locus allele, *A3*). Additional markers at intervals along the haplotype would add greatly to our ability to map the histocompatibility and disease-associated loci and thus to investigate the nature of their products and to add to predictive value. Fortunately many additional markers are being discovered. These include biochemical markers, such as electrophoretic variants of glyoxalase (GLO) and pepsinogen 5 (Pg 5) or complement (C6), red cell antigens Chido and Rogers, and a host of new glycoproteins found only on B (surface immunoglobulin bearing) lymphocytes (B cell antigens).

Although antigens expressed predominantly on B cell have been identified for several years in the mouse as Ia antigens, realization came late that similar antigens are present in man. This is because HLA typing is performed mainly on peripheral blood lymphocytes (PBL) and only a minority, usually 5–15%, of PBL are B cells. Any reactions that were observed were dismissed as background or uninterpretable weak and sporadic cross-reactions. With the realization of a new universe of B cell specificities, procedures were rapidly developed for the separation of B lymphocytes from T lymphocytes and monocytes, while reagent antisera could be obtained from existing anti-HLA sera by removal of anti-HLA activity. This can be accomplished by mixing the serum with platelets. Platelets carry HLA, but not B cell antigens, and hence the anti-HLA antibody combines with the platelet which can be mechanically removed and leaves the desired anti-B cell activity in the supernatant (absorbed) serum.

One B cell locus lies very close to HLA-D and has many alleles.

The existence of other B cell loci near D and possibly near A is inferred, but these loci are not yet mapped. Although for technical reasons B cell typing is by no means routine, a widespread application is likely in the near future.

Reference has been made to HLA-D without explaining what it is. HLA-D is the name given to a locus on the HLA haplotype that controls the ability of a B lymphocyte to stimulate a responding T cell from a person of different HLA-D type in the mixed lymphocyte reaction (MLR). Initially MLR was informative only within a family. Because HLA is inherited as a unit, there are usually only 6 HLA combinations in a two-generation family. The paternal haplotypes are usually designated a and b, the maternal c and d. The children are then ac, ad, bc, or bd. Where there are five children at least two of them must be HLA identical. Cells from such pairs do not stimulate; virtually all other combinations within the family or between unrelated people do stimulate. The important exceptions are stimulating cells from individuals having the same HLA-D allele on each haplotype. Such individuals are rare; many of the 100 or so now identified are children of first cousin marriages or are from inbreeding groups. The cells are called HLA-D homozygous typing cells (HTC), and five alleles have been officially recognized (Table 1). If HLA-Dw1 HTC are mixed with lymphocytes from a responder who has the Dw1 allele, no stimulation is observed. Dw1 HTC will of course stimulate a donor who is, for instance, *Dw2/Dw3.* The MLR is time consuming (5–7 days) to set up and results are greatly influenced by environmental variation (pH and temperature, source of support serum used, etc.). It is very encouraging therefore to find that there is an excellent correlation between the reactivity of some anti-B cell sera and the distribution of HLA-D. Because the association with disease is generally higher with HLA-D than with HLA-B, screening individuals at risk for a predisposition to some diseases, e.g., children in families where other members have juvenile onset diabetes, multiple sclerosis, leprosy, or other HLA associated diseases, may be accomplished more quickly and conveniently by testing for B cell antigens. It is possible that some of the B cell antibodies will be directed toward receptors on the cell surface, thus helping in the study of the etiology of disease as well as in prophylaxis.

Many populations have been tested and about 40 diseases have now been identified as having an HLA association (11). The relative risk is in many instances rather low and with the exception of AS, HLA typing is of limited value to the clinician except in families known to have a genetic predisposition to disease. Nonetheless genetic studies are very encouraging. Several groups of investigators had tried to link leprosy to HLA, but none had been strikingly successful until de Vries and his colleagues (12) found that in families having members with both lepromatous and tuberculoid leprosy, the form of the disease correlated with the distribution of HLA haplotypes within each of the families studied. Since patients with lepromatous leprosy have impaired immune function it is tempting to speculate that a gene associated with HLA is controlling the level of immunological responsiveness in these families.

Throughout this chapter the emphasis has been on the HLA antigens as markers, and this is what they are best regarded as, at present. The actual function of the HLA molecules is unknown. In the mouse the H-2 antigens and the Ia antigens (analogues of the human B cell antigens) serve in cell-cell collaboration in the immune response, and it is probable they play a similar role in man.

ACKNOWLEDGMENTS

I would like to thank D. Kostyu and K. Singer for critical review of the manuscript and J. Kerber for her help in its preparation.

LITERATURE CITED

1. Terhorst, C., Robb, R., Jones, C., and Strominger, J. L. Further structural studies of the heavy chain of HLA antigens and its similarity to immunoglobulins. Proc. Nat. Acad. Sci. In press.
2. Parham, P., Albert, B. N., Orr, H. T., and Strominger, J. L. The carbohydrate moiety of HLA antigens: Structure, antigenic properties and amino acid sequences around the site of glycosylation. J. Biol. Chem. In press.
3. Kissmeyer-Nielson, F., and Kjerbye, K. E. 1967. Lymphocytotoxic microtechnique purification of lymphocytes by flotation. In Histocompatibility Testing, pp. 381–383. Munksgaard, Copenhagen.
4. Terasaki, P., McClelland, J. D., Park, M. S., and McCurdy, B. Microdroplet lymphocyte cytotoxicity test. In Manual of Tissue Typing Techniques, DHEW publ. (NIH) 74-545. Government Printing Office, Washington, D.C.
5. Amos, D. B., and Pool, P. 1976. HLA typing. In N. R. Rose and H. Friedman (eds.), Manual of Clinical Immunology, pp. 797–804. American Society for Microbiology, Washington, D.C.
6. van Rood, J. J. 1963. Leucocyte grouping. A method and its application. Ph.D. dissertation, University of Leyden, Netherlands.
7. Selwood, N. H. 1975. Detailed numerical analysis of serological data from the Fifth International Histocompatibility Workshop. Tissue Antigens 5:367–385.
8. Ward, F. E., de Jongh, D. A., and Biefel, A. A. HLA antigens in North American Blacks. In Histocompatibility Testing. In press.

9. Yunis, E. J., and Amos, D. B. 1971. Three closely linked genetic systems relevant to transplantation. Proc. Nat. Acad. Sci. USA 68:3031-3035.
10. Long, M. A., Handwerger, B. S., Amos, D. B., and Yunis, E. J. 1976. The genetics of cell-mediated lympholysis. J. Immunol. 117:2092-2099.
11. Svejgaard, A., Jersild, C., Nielsen, L. S., and Bodmer, W. F. 1974. HL-A antigens and disease. Statistical and genetical considerations. Tissue Antigens 4:95-105.
12. de Vries, R. R. P., Lai A Fat, R. F. M., Nijenhuis, L. E., and van Rood, J. J. 1976. HLA linked genetic control of the host response to Mycobacterium leprae. Lancet 2 (7999):1328-1330.

Infection, Immunity, and Genetics
Edited by Herman Friedman, T. Juhani Linna, and James E. Prier

MACROPHAGE FUNCTION IN GENETIC CONTROL OF THE IMMUNE RESPONSE

Alan S. Rosenthal,
Lanny J. Rosenwasser,
and Joseph T. Blake

BACKGROUND INFORMATION

It is generally recognized that macrophages facilitate a variety of in vitro thymic-dependent immunological phenomena in the mouse, guinea pig, and man (1). The cellular and molecular events that underlie their function have been analyzed. Antigen-specific activation of mediator production and of clonal expansion of T lymphocytes requires the presence of macrophages, and the effects of macrophages in such activation extend beyond maintenance of in vitro culture conditions (2,3). Indeed, in the guinea pig one can clearly demonstrate that recognition of soluble protein antigens by the T lymphocyte requires an initial uptake of antigen by the macrophages (3). These characteristics are summarized in Table 1. The biologic constraints inherent in a two-cell immunogenic recognition unit are clear. It is necessary to define both the cellular and immunological specificity of this requirement for macrophages. In the simplest experiments, the proliferative capacity of populations of lymphoid cells depleted of macrophages are restored by adherent cells but not by lymphocytes, granulocytes, or fibroblasts (2). This requirement for macrophages in T cell antigen recognition is not dependent on the immune state of the animal from which the macrophage was obtained, although macrophages and lymphocytes must share genetic identity at some portion of the major histocompatibility complex (MHC) for successful detection of the antigenic signal borne by the macrophage (4) (Table 2).

The cellular requirements for murine T cell antigen recognition are less well understood and were therefore evaluated in an adherent-

Table 1. Macrophage function in recognition of antigenic signals by T lymphocytes

1. Absolute macrophage requirement exists for induction of proliferation and helper cell activity.
2. Antigens are initially bound and internalized by macrophages.
3. The physical state of macrophage-associated antigen is ordered.
4. Delivery of macrophage "processed" antigen signals occurs by either direct physical interaction or by release of antigen-self complexes.
5. Macrophage-T lymphocyte interactions are structured by products of genes linked to the major histocompatibility complex of the species, i.e., antigen recognition involves some form of self-recognition.

cell dependent in vitro assay. The immune response of sensitized T cells to complex multideterminant soluble protein antigens such as dinitrophenylated ovalbumin is dependent on the production of a soluble factor that is not antigen specific nor H-2 related in that adherent cell, regardless of haplotype produce factor independent of the presence of antigen. Mouse lymphocytes do not, however, respond to supernates of either guinea pig macrophages or fibroblast, indicating the existence of species and cell type specificity. By contrast, the adherent cell requirement for T cell proliferation to the synthetic terpolymer L-glutamic acid, L-lysine, L-tyrosine (GLT[15]), an antigen, the response to which is under specific *Ir* gene control, necessitated identity or partial identity at the major histocompatibility complex between macrophage and primed T cell (5).

Is macrophage lymphocyte interaction a general requirement for all T cell activation or do these associations function solely in the recognition of cell-bound and soluble protein antigens? The response of T cells to phytohemagglutinin (PHA) requires accessory cells; but

Table 2. Capacity of macrophage-bound antigen to activate lymphocyte proliferation

	Lymphocyte		
Macrophage strain	Strain 2	Strain 13	F_1 hybrid (2 × 13)
2	Yes	No	Yes
13	No	Yes	Yes

unlike the macrophage requirement for soluble protein antigen recognition, their function can be replaced by fibroblasts or increased cell density (6).

It is crucial therefore to establish the nature of antigen uptake processes and the physical state of macrophage-associated, metabolized antigen (7).

Cytochalasin B, an agent active against a variety of membrane functions that also blocks pinocytosis, does not affect uptake of immunologically relevant protein. Trypsinization of macrophages pulsed with antigen at 4 °C removes all immunological activity without altering the ability of such cells to take up new antigens. If, however, trypsinization is delayed by prior culture, pulsed macrophages become progressively resistant to the effects of trypsinization such that by 120 min the immunologic potential of the macrophages is unaltered.

Macrophage-lymphocyte interaction may occur either indirectly via the secretion of antigen complexed to soluble macrophage factors cytophilic for lymphocytes or via direct cell interaction. We have found that at least two types of physical interactions take place. The first occurs rapidly, is independent of the presence of antigens, and is without immunologic commitment (8). This step requires active macrophage metabolism, divalent cations, and involves a trypsin sensitive macrophage site. This binding does not distinguish T or B cells and is reversible such that at any given time binding represents an equilibrium between cellular association and dissociation. When this step has brought specifically immune lymphocytes into apposition with antigen bearing macrophages, a second type of binding results (9). This latter phenomenon is dependent upon the presence of antigen and a sharing on lymphocyte and macrophage of histocompatibility-linked gene products. This association is not easily reversed and eventuates in proliferation of the bound lymphocyte (Figure 1).

The use of antigens under control of dominant H-linked genes and specific alloantisera that recognize cell-associated antigenic determinants, which are also products of H-linked genes, provides useful biological probes of the function and cellular location of molecules; these play critical but incompletely defined roles in the T lymphocyte antigen selection or activation process. Alloantisera blocked the antigen response controlled by an *Ir* gene linked to the gene coding for the alloantigen, but had little effect on the antigen response controlled by an *Ir* gene that is the product of the opposite haplotype (10). The inhibitory activity of the alloantisera was initially presumed to be directed solely against the proliferating T lymphocyte. However, the

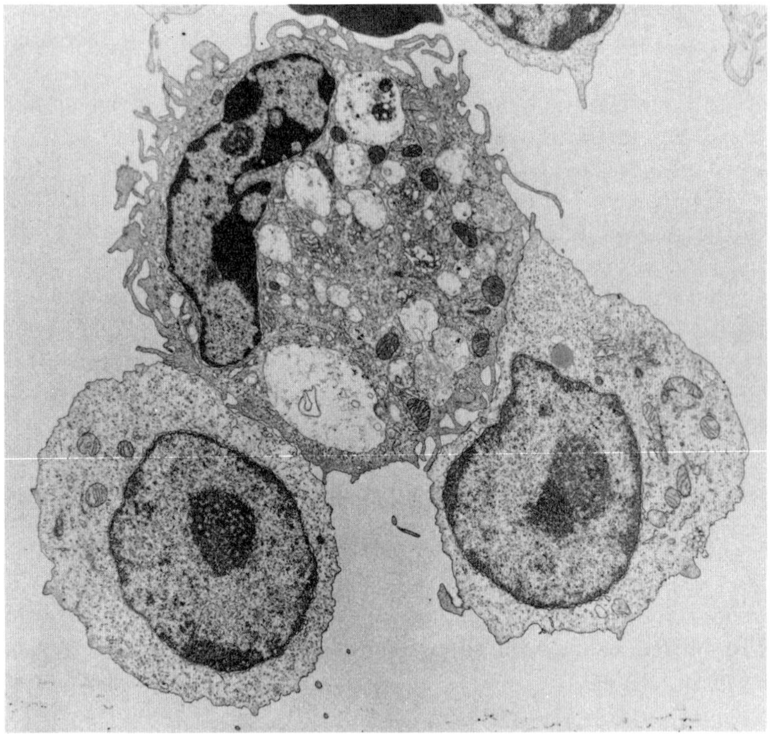

Figure 1.

central role of the macrophage in the activation of T-cell proliferation suggests that an alternative site for the inhibitory effect of the alloantisera might be on the macrophage itself by blockade of the physical and/or functional interaction of the macrophage and T cell. Unfortunately, experiments that attempt to define the cellular site of action of the alloantisera using combinations of antigen-pulsed macrophages and T cells are not possible because efficient interaction of macrophage-associated antigen and T cell is only seen when the macrophage and T cell share identity of gene products linked to the guinea pig MHC.

There are, however, a number of special conditions in which a macrophage may "paradoxically" induce allogeneic lymphocyte proliferation. These circumstances are revealing situations in that they clearly suggest that the failure to recognize soluble protein antigens bound to allogeneic macrophages is not a primary inability of such cells to present signals but may instead reflect a basic feature of antigen handling by macrophages.

Guinea pig macrophages are very effective stimulators of allogeneic lymphocyte proliferation (11). Thus, soluble proteins may be displayed in an immunologically relevant fashion only on syngeneic macrophages, whereas in the MLR it is the histoincompatible macrophage that induces T cell proliferation, again suggesting that the critical difference lies in the display of alloantigen as opposed to that of foreign proteins.

In another model system, chemical alteration of the macrophage-depleted lymphocyte or macrophage plasma membrane glycoproteins by treatment with either sodium metaperiodate or neuraminidase galactose oxidase generates free aldehyde groups on terminal sugar moieties (12). Subsequent and rather massive lymphocyte proliferation will occur if and only if modified lymphocyte is combined with unmodified macrophages or modified macrophages are combined with unmodified lymphocytes. Moreover, proliferation will occur irrespective of the histocompatibility difference between the macrophage and lymphocyte. These latter observations have led us to suggest that immune recognition may in fact consist of at least two separate but linked events. The first phase of this interaction involves the selective binding of antigen specific T cells by macrophages bearing that antigen. The second or "activation" phase of macrophage lymphocyte interaction is a relatively non-specific event that follows as a consequence of stabilization of physical macrophage lymphocyte interaction initiated by the antigen-specific selection process.

Models for the role of the macrophage in genetic control of T cell responses to soluble protein antigens must consider: 1) an antigen specific receptor on the T cell, 2) cellular interaction structures identical to or closely linked to the alloantigens, 3) the *Ir* gene product functioning either as the specific T cell receptor or as an auxiliary antigen recognition molecule on the macrophage and/or T cell, 4) antigen bound to metabolically intact macrophages in a form resistant to proteolytic treatment and located in a cellular site not available to specific antibody, and 5) a macrophage-associated non-specific membrane trigger site or secretory product as suggested by the activation of aldehyde-treated lymphocytes by normal macrophages.

At present we favor the hypothesis that states that a given macrophage's repertoire of *Ir* gene products functions to specify determinant selection. Possible mechanisms by which the interaction of antigen with the *Ir* gene product could select determinant specificity are:

1. *Enzymatic.* The *Ir* gene product is an enzyme of relative broad specificity, such as a protease, which destroys only those polypeptide sequences toward which it has reactivity.
2. *Conformation.* The *Ir* gene product alters the tertiary structure of the antigen itself so as to display selected regions of the polypeptide to the T cell.
3. *Receptor presentation.* This would require that certain similar existing amino acid sequences or conformations in a complex antigen be recognized by *Ir* gene products. Such receptor molecules would possess specific binding properties akin to those of the binding site of a proteolytic enzyme but would differ from the enzymatic model in that no selective degradation of the antigen would occur. The function of such class-specific antigen binding molecules would be to orient or to display those regions of the polypeptide antigen not bound to the receptor.

GENETIC CONTROL OF THE IMMUNE RESPONSE TO INSULIN IN THE GUINEA PIG

Genetic control of immune responsiveness to both synthetic and natural polypeptide antigens has been found in experimental animals. Unfortunately absolute demonstration of such phenomena in man has not been possible because of the limited number of ethical antigens that can be used in man and their lack of defined physical structure. One antigen that might be of significant potential in such studies is insulin. Beginning with studies in inbred Strains 2 and 13, we have found that cellular and humoral immune responsiveness of distinct regions of the molecule are under control of histocompatibility-linked genes (13). Strain 2 guinea pig T cells (as measured by proliferation or T cell help) recognize a determinant consisting of amino acids 6–11 of the A chain (α loop), whereas Strain 13 guinea pig T cells recognize a region in the insulin B chain. Nonetheless the determinants recognized by antibody generated in both strains predominantly recognize shared determinants.

DETERMINANT SELECTION: A MACROPHAGE *Ir* GENE FUNCTION

Although immune specificity is traditionally considered a lymphocyte function, observations from the above as well as from other studies suggest an interpretation that the macrophage has some discriminatory function in the antigen recognition process (Table 3). We thus assessed the function of macrophages in *Ir* control, taking advantage of the capacity of T cells from insulin-immunized F_1 (2 × 13) guinea pigs to respond to antigen-bearing parental macrophages (14). We reasoned:

If *Ir* gene function operates exclusively at the level of the responding lymphocyte, both determinants should be simultaneously recognized by F_1 T cells independently of which parental macrophage is "presenting" the antigen. Conversely, if the definition of which determinant is actually recognized depends on the genetic profile of the antigen-presenting macrophage, then *Ir* gene function must also be operating at this cell level. The data favor the latter assumption. Indeed our studies of the immune response to insulin raise the possibility that a selected amino acid sequence and/or conformation within the antigen itself is seen by the T cell receptor. Generation or display of such antigenic determinants would thus be a function of immune response genes operating at the level of the antigen presenting cell. Two general mechanisms by which macrophages might subserve such a function may be proposed. One possibility is that immune response genes define a class of receptors or broad specificity which recognize molecular shape and thus have the unique ability to focus or orient distinct regions of the antigen for presentation to the T cell. A second

Table 3. Data that suggest that the macrophage has discriminatory function in genetic control of antigen recognition

1. The response of F_1 (2 × 13) guinea pigs primed to the synthetic random co-polymers, GL (L-glutamic acid, L-lysine) and GT (L-glutamic acid, L-tyrosine) is under control of MHC-linked *Ir* genes. T lymphocytes from such animals efficiently recognize antigen bound only to parental maccrophages of responder origin (Strain 2 for GL, and Strain 13 for GT) while macrophage from both parental strains can "present," to F_1 T cells, antigen not under unigenic control.
2. Alloantisera recognizing MHC-linked cell surface determinants, *Ia* antigens, are able to block T cell proliferation to antigen-bearing, chemically modified, and allogeneic macrophages. In the latter two situations, DNA synthesis can be initiated across an MHC barrier and thus allowed definition of the site of alloantisera blockade. Such experiments show that the inhibitory activity of alloantisera is directed against the *Ia* specificities of the participating macrophage and not to those of the responding lymphocyte.
3. Studies in mice and guinea pigs have shown that secondary T cell responses are elicited not against the intact priming antigen alone but rather to antigen in the context of MHC-linked cell surface structures present on the original antigen presenting cell.

possibility is that immune response gene products are, or regulate the activity of, families of enzymes that modify or metabolize polypeptide antigens. In this latter situation, the repertoire of *Ir* genes associated with a given haplotype would define restricted areas of the molecule as available for display to the T cell receptor. Experiments to distinguish these possibilities are in progress.

The characterization of the immune responsiveness of thymus-derived lymphocytes provides useful models for study of the role of the cellular immune effector cells in immunologic surveillance and their contribution to the humoral limb of host defenses. Microbial immunity is known to rely to a great extent upon mechanisms of T cell-mediated cellular immune function such as enhanced bacterial killing mediated by the elaboration of so-called "soluble mediators." A more fundamental understanding of the immunobiology of these crucial events may provide new insights into immune competence in man and information on the mechanisms of cellular regulation of immune responsiveness.

LITERATURE CITED

1. Rosenthal, A. S., and Shevach, E. M. 1976. Contemporary Topics in Immunobiol. 5:47.
2. Rosenstreich, D. L., and Rosenthal, A. S. 1974. J. Immunol. 112:1085.
3. Waldron, J. A., Horn, R. G., and Rosenthal, A. S. 1973. J. Immunol. 111:58.
4. Rosenthal, A. S., and Shevach, E. M. 1973. J. Exp. Med. 138:1194.
5. Rosenwasser, L., and Rosenthal, A. S. J. Immunol. In press.
6. Lipsky, P. E., and Rosenthal, A. S. 1975. J. Immunol. 115:440.
7. Ellner, J. J., and Rosenthal, A. S. 1975. J. Immunol. 114:1563.
8. Lipsky, P. E., and Rosenthal, A. S. 1973. J. Exp. Med. 138:900.
9. Lipsky, P. E., and Rosenthal, A. S. 1975. J. Exp. Med. 141:138.
10. Shevach, E. M., and Rosenthal, A. S. 1973. J. Exp. Med. 138:1213.
11. Greineder, D. K., and Rosenthal, A. S. 1975. J. Immunol. 115:932.
12. Greineder, D. K., and Rosenthal, A. S. 1975. J. Immunol. 114:1541.
13. Barcinski, M. A., and Rosenthal, A. S. 1977. J. Exp. Med. 145:726.
14. Rosenthal, A. S., Barcinski, M. A., and Blake, J. T. 1977. Nature (London) 267:156.

Infection, Immunity, and Genetics
Edited by Herman Friedman, T. Juhani Linna,
and James E. Prier

GENETIC CONTROL OF IMMUNITY TO BACTERIAL ANTIGENS— AN INTRODUCTION

Toby K. Eisenstein

In John Gowen's 1948 review on "Inheritance of Immunity in Animals," he compiled the literature on the selective resistance of some members of a species to infection with a given pathogen (1). The major conclusion that could be drawn from the data available at the time was that there is a genetic basis for susceptibility and resistance to many bacterial, viral, and protozoan infections, which can be differentiated from environmental factors, such as diet, or previous exposure of the host to the infectious agent. The evidence for this conclusion rested principally on breeding experiments, in which strains of resistant and susceptible animals were developed. Webster in the 1930s bred two lines of mice, designated BSVS and BRVR, which were susceptible or resistant to murine salmonellosis or St. Louis encephalitis (B = Bacillus, V = Virus, S = Susceptible, R = Resistant) (2). Through appropriate crosses, offspring were obtained that were susceptible to one agent and resistant to the other, susceptible to both, or resistant to both, showing that the resistance to these two infectious agents was under separate control. In later experiments, Hill, Hatswell, and Topley bred a line of mice that was markedly endotoxin resistant but was *Salmonella* susceptible, showing that these traits may not necessarily be related (3). In the 1950s Max Lurie, while at the Phipps Institute, developed a colony of rabbits from which he selected animals that were susceptible or resistant to tuberculosis infection (4). (As an interesting aside, he was a member of the Eastern Branch of the Society of American Bacteriologists, the forerunner of the sponsoring organization of this symposium. I am told that he frequently and enthusiastically reported his findings at the branch meetings.)

Over the years, attempts have been made to discover the mechanisms behind the observed differences in susceptibility of various inbred strains. In 1923 Rich showed that susceptibility of a strain of guinea pig to *S. choleraesuis* infection was caused by a heritable deficiency of a component of the complement system (1). Until recently, this stood as one of the few cases where a single factor could be identified as the cause of genetic differential susceptibility to infection. In 1976, Miller et al. reported that human susceptibility to *Plasmodium vivax* infection correlates with the presence of the Duffy blood group on the red blood cells of the host (5).

Nevertheless, in many other systems, in spite of extensive experimentation, no single factor has emerged to explain the observed differences. The BSVS and BRVR mice have been compared for rates of blood clearance of *Salmonella,* for serum opsonin levels, and for complement levels, but no significant differences in these parameters have been found (6, 7). Lurie initially thought that hormones might be the important factors in the anti-tuberculosis resistance of his rabbits, and he carried out extensive experiments in vivo to correlate levels of various hormones with resistance to infection. Later he concluded that the critical factor for host defenses against mycobacteria was the macrophage (4). Maier and Oels, in 1972, provided evidence that macrophages were also involved in the differential susceptibility of BSVS and BRVR mice to *S. typhimurium,* another intracellular pathogen. They found that the macrophages from BRVR mice were more bactericidal for *Salmonella* than those for BSVS mice (8).

The theory that differences in susceptibility and resistance to some intracellular pathogens are caused by differences in the bacteriostatic or bactericidal power of host macrophages is widely accepted. One should be cautious about attributing too much significance to in vitro analyses of single factor differences, however, because it may be difficult to assess the relative importance of such differences for host defenses in vivo. Thus, Swartzberg, Krahenbuhl, and Remington showed that macrophages taken from mice given *Toxoplasma gondii* or *Corynebacterium parvum* in vivo were activated and were able to kill *T. gondii* in vitro. However, when *C. parvum*-treated mice were challenged with *T. gondii* in vivo, they were not protected, but *Toxoplasma*-immunized mice were resistant (9). Additional evidence that cautions against generalization about mechanisms of resistance to intracellular pathogens is also presented by Dr. Vas. He shows that, although the macrophage is thought to be central to resistance against both *Listeria* and *Salmonella* infections, strains of mice are available that are resistant to one organism but susceptible to the other. These types of studies indicate that there are either quantitative or qualitative differences in macrophage activation, or that there are other parameters involved in host resistance to these microorganisms.

The papers that have been presented in the first part of this volume show that the ability of animals to give an immune response to some antigens is linked to the major histocompatibility locus. The following papers concern studies on the genetics of the immune response to pneumococcal polysaccharide and to the lipopolysaccharide (LPS) of Gram-negative bacteria, as well as on the genetics of susceptibility to *Salmonella* and *Vibrio cholerae* infection. In none of these murine model systems has a correlation been found between the H-2 locus and resistance to infection, or between the H-2 locus and the ability to respond immunologically to the bacterial antigen. In the case of pneumococcal polysaccharide, Baker et al. show that responsiveness is under polygenic control and that the genes are not the same as the one involved in responsiveness to LPS. In the case of lipopolysaccharide, Watson, Kelly, and Largen present evidence for a single autosomal gene controlling susceptibility to LPS's toxic effects and discuss the progress being made toward mapping that gene. In this system we are approaching the threshold of defining not only the genetic basis but also the molecular basis for toxicity of this important bacterial product. Drs. Friedman and Baker, working in the cholera and pneumococcal systems, respectively, show how recent knowledge about triggering and regulation of the immune response is being applied to further our understanding of the cellular and molecular basis for the genetic differences in host susceptibility and immune responsiveness to these organisms and their products.

LITERATURE CITED

1. Gowen, J. W. 1948. Inheritance of immunity in animals. Annu. Rev. Micro. 11:215–254.
2. Webster, L. T. 1937. Inheritance of resistance of mice to enteric bacterial and neurotropic virus infections. J. Exp. Med. 65:261–286.
3. Hill, A. B., J. M. Hatswell, W. W. C. Topley. 1940. The inheritance of resistance, demonstrated by the development of a strain of mice resistant to experimental inoculation with bacterial endotoxin. J. Hyg. 40: 538–547.

4. Lurie, M. 1964. Resistance to Tuberculosis: Experimental Studies in Native and Acquired Defensive Mechanisms. Harvard University Press, Cambridge.
5. Miller, L. H., S. J. Mason, D. F. Clyde, and M. H. McGinniss. 1976. The resistance factor to *Plasmadium vivax* in Blacks. New England J. Med. 295:302-304.
6. Böhme, D. H., H. A. Schneider, and J. M. Lee. 1959. Some pathophysiological parameters of natural resistance in murine salmonellosis. J. Exp. Med. 110:9-25.
7. Gröschel, D., C. M. S. Paas, and B. S. Rosenberg. 1970. Inherited resistance and mouse typhoid. J. Reticuloend. Soc. 7:484-499.
8. Maier, T., and H. C. Oels. 1972. Role of the macrophage in natural resistance to Salmonellosis in mice. Infect. Immun. 6:438-443.
9. Swartzberg, J. E., J. L. Krahenbuhl, and J. S. Remington. 1975. Dichotomy between macrophage activation and degree of protection against *Listeria monocytogenes* and *Toxoplasma gondii* in mice stimulated with *Corynebacterium parvum*. Infect. Immun. 12:1037-1043.

Infection, Immunity, and Genetics
Edited by Herman Friedman, T. Juhani Linna, and James E. Prier

THE GENETIC MAPPING OF A LOCUS IN MICE THAT CONTROLS IMMUNE AND NONIMMUNE RESPONSES TO BACTERIAL LIPOPOLYSACCHARIDES

James Watson, Kathleen Kelly, and Michael Largen

Lipopolysaccharides (LPS) isolated from Gram-negative bacteria possess many diverse biological properties. The immunological responses elicited by LPS have been of particular interest for a number of reasons. LPS are extremely potent antigens; small doses elicit very large antibody responses directed against the *o*-specific antigens (1, 2). In addition LPS seem to non-specifically stimulate the immune system, enhancing humoral responses to a variety of antigens (3, 4). The characterization of both the structural and the biological properties of LPS has revealed that the immune system responds to two distinct moieties: the *o*-polysaccharide as the major antigenic structure and lipid A which is responsible for the adjuvant effects (5). It is apparent that the immune system is not unique in its response to lipid A. The bewildering array of endotoxic reactions elicited in mammals by LPS all seem to be initiated by the lipid A moiety. The studies in these authors' laboratory on the genetic control of responses to LPS in

---

Supported by Public Health Service Grant AI 13383 from the National Institute of Allergy and Infectious Diseases and Grant 1-469 from the National Foundation. James Watson is supported by a Research Career Developmental Award AI-00182 from the National Institute of Allergy and Infectious Diseases, Kathleen Kelly is supported by NCI Training Grant CA 09054, and Michael Largen is supported by a National Institute of Health Training Grant GM 07307. We thank Dr. B. Taylor and Dr. K. McAdam for their collaborative efforts in these studies.

mice are summarized herein. First the genetic control of various responses of the immune system to lipid A, and then several nonimmune responses to lipid A are discussed. It is suggested that most biological responses elicited by LPS result from the interaction of lipid A with receptors that may be distributed on many different cell types. How the results of genetic analyses may relate to an understanding of the endotoxic effects of LPS that are apparent during infections caused by Gram-negative bacilli is also covered.

RESPONSE OF THE IMMUNE SYSTEM TO LPS

Lipid A acts as a specific mitogen for bone marrow-derived (B) lymphocytes in mice (3, 6, 7). This results in the polyclonal expression of antibody synthesis (7). In addition lipid A acts as an adjuvant for the specific antibody response to the *o*-antigens of the polysaccharide moiety in LPS, stimulating a larger immune response than observed to the polysaccharide antigen alone (8). The adjuvant activity of lipid A is also assayed by the enhancement of immune responses to soluble antigens such as bovine serum albumin (3, 4) and the capacity to modulate the induction of a specific state of tolerance to several thymus-dependent antigens into a specific state of immunity (9).

A variety of structural and functional criteria have been used to divide B lymphocytes into subpopulations. These include the expression of surface immunoglobulin (Ig) and the antigens associated with the immune response genes (Ia) (10, 12). LPS is mitogenic for some but not all subpopulations of B lymphocytes. However, LPS will induce a number of nonproliferative changes in other subpopulations of B lymphocytes. In particular, LPS will induce subpopulations of immature B lymphocytes to express surface Ig, Ia, and the receptor for the third complement component (CR) (10–12). These inducible cells are found in the bone marrow or spleen of mice, and it has been suggested the surface antigens are expressed in a specific temporal sequence that reflects the changing phenotypes of cells differentiating through the B lymphocyte lineage (12).

The C3H/HeJ mouse strain differs in responsiveness to LPS when compared with other strains of mice, and as a result has become an important tool for analyzing the genetic control of responses to LPS. C3H/HeJ mice are refractive to the mitogenic (13, 14) and polyclonal effects of LPS (15). This defect seems specific in that other B cell mitogens such as polyinosinic acid, dextran sulphate, and a purified protein derivative from tuberculin stimulate mitogenic responses in this strain. It is important to separate in C3H/HeJ mice the responses to the two different regions of LPS. Immune responses to LPS are restricted to only those B lymphocytes that possess surface Ig receptors that bind the *o*-antigenic determinants. C3H/HeJ mice support immune responses to the polysaccharide moiety of LPS; thus these antigen sensitive cells are present. However, all lipid A-induced responses seem defective (16), including adjuvant effects which are apparent in the immune response to *o*-polysaccharide antigens of LPS itself as well as to soluble proteins (3, 16). Utilizing the assay system described by Komuro and Boyse (17), it has also been shown that lipid A does not induce changes in the immature B lymphocyte subpopulation from C3H/HeJ mice (18). Thus, lipid A fails to elicit either the nonproliferative or the proliferative responses in subpopulations of murine B lymphocytes.

GENETICS OF DEFECTIVE LPS RESPONSES IN C3H/HeJ MICE

Early studies demonstrated that the F_1 hybrids (C3H.SW-Ig-1^b × C3H/HeJ) exhibited a mitogenic response to LPS that was as high as the C3H.SW-Ig-1^b (CWB/13) parent (14). This argued for dominant expression of LPS responsiveness. We have now examined F_1 hybrid mice between C3H/HeJ mice and a number of LPS responder strains and all show intermediate responses to LPS (19). The response of CWB/13 mice to LPS is comparable to that observed in the other LPS responder parent strains that were investigated. At present we have no adequate explanation for the dominant expression of LPS responsiveness in (C3H/HeJ × CWB/13) F_1 hybrids.

The intermediate response of F_1 animals was examined further by considering the kinetic patterns of mitogenic responsiveness. At a cell density of 5.0×10^6 cells/ml, the kinetic patterns of LPS responsiveness differ in F_1 hybrid and responder parent spleen cell cultures. Parent cultures show a sharp peak of mitogenic responsiveness on day 2, whereas F_1 cultures show a broad plateau of maximal responsiveness on days 2 and 3 (19). A variable that seems to influence kinetic patterns is the density of cells in a culture. When the density of cells in a responder parent culture is reduced to 1.25×10^6 cells/ml, the kinetic pattern of mitogenic responsiveness is similar to that observed in F_1 cultures at a cell density of 5.0×10^6 cells/ml or 2.5×10^6 cells/ml. Autoradiographic studies have indicated that the number of LPS responsive cells in F_1 cultures is approximately half the number of responsive cells in parent cultures. These data correlate with the kinetic patterns in the two types of cultures and imply that the number of responsive cells in a culture is an important parameter in determining the kinetic pattern. Thus, it seems that kinetic responses to LPS most likely are primarily a result of the culture conditions and are not directly under genetic control (19).

In order to determine the number of genes involved in the defective LPS response in C3H/HeJ mice, we have performed a number of backcross linkage analyses. In backcross progeny from three F_1 strains of mice, (C3H/HeJ × CWB/13) F_1, (C3H/HeJ × C57BL/6J) F_1 and (C3H/HeJ × BALB/c) F_1, each backcrossed to the C3H/HeJ parent; responder and nonresponder progeny were observed in a ratio of 1:1 (14, 19). These results suggest that the lack of LPS responsiveness in C3H/HeJ mice is controlled by a single locus, and the alleles at this locus are co-dominantly expressed (19). Our data are consistent with the F_1 studies and the backcross and F_2 segregation analyses that have been published by other laboratories (20, 21, 22).

The intermediate mitogenic responsiveness of F_1 cultures results from approximately half the number of LPS responsive cells in F_1 cultures as compared to the number of responsive cells in responder parent cultures. This finding suggests either a gene dosage phenomenon or allelic exclusion at the locus that controls LPS responsiveness (19). The resolution of these alternatives requires identification of the LPS response locus gene product and the ability to assay individual cells for the presence of the known gene product.

We have also compared the linkage relationships of the polyclonal and adjuvant responses to mitogenic responses in backcross (C3H/HeJ × CWB/13) F_1 × C3H/HeJ mice (14, 15, 23). Mitogenic, polyclonal, and adjuvant responsiveness to LPS all segregated together in the backcross progeny, demonstrating the expression of a single gene controlling each of these responses to LPS (14, 15, 23).

Each of these responses is an assay of the expression of B lymphocytes. The defective LPS responses in C3H/HeJ mice may involve a gene expressed in B lymphocytes or in some other cell type. We have performed a variety of in vivo and in vitro cell reconstitution experiments and these all indicate that the defect in C3H/HeJ mice that restricts mitogenic responsiveness is expressed in the B lymphocytes (15).

GENETIC MAPPING OF THE DEFECTIVE LPS RESPONSE GENE

The use of a number of recombinant inbred strains of mice has enabled us to extend our genetic studies of the defective LPS response

Table 1. Mitogenic and adjuvant responses to LPS in BXH recombinant inbred strains[a]

Strain	Mitogenic response (Cpm × 10^{-3})[b]		Serum titers[c]	LPS response
	No LPS	25 μg		
C3H/HeJ	4.4	5.6	<10/10	Low
C57BL/6J	3.8	51.5	<10/1,280	High
BXH-2	1.1	5.1	<10/10	Low
BXH-3	1.1	3.1	<10/20	Low
BXH-4	3.7	8.3	10/20	Low
BXH-5	3.7	50.8	<10/1,280	High
BXH-6	3.3	4.5	<10/40	Low
BXH-7	2.2	3.0	10/80	Low
BXH-8	1.2	2.5	<10/10	Low
BXH-9	3.7	5.3	<10/10	Low
BXH-10	3.3	8.1	10/80	Low
BXH-11	2.2	43.1	10/320	High
BXH-12	2.8	5.6	<10/10	Low
BXH-14	3.9	46.6	10/1,280	High
BXH-18	3.1	5.4	10/80	Low
BXH-19	3.5	44.5	10/320	High

[a]The data are summarized from Reference 24.

[b]Mitogenic assays were performed as detailed in Reference 24.

[c]Two figures are presented for the serum titers. The first represents the titer of the pre-bled serum, and the second represents the titer 7 days after immunization. The serum titers are presented as a reciprocal of the end point dilution. The initial dilution was always 1:10.

gene in C3H/HeJ mice. Fourteen RI strains have been produced by inbreeding, beginning with randomly chosen pairs of mice, from the F_2 generation of the cross between C3H/HeJ and C57BL/6J progenitor strains (24). The data presented in Table 1 describe the mitogenic response of each of the BXH strains to various concentrations of LPS. Spleen cultures were prepared from mice from each of the 14 BXH strains and were incubated with various concentrations of *Escherichia coli* 0127:B8 LPS. Spleen cultures from the BXH-5, 11, 14, and 19 strains all showed mitogenic responses to LPS quantitatively similar to those observed in cultures prepared from C57BL/6J mice. The BXH-10 strain initially showed low mitogenic responses, and the remaining BXH strains showed no mitogenic responses to LPS. Similar results have been obtained using *E. coli* K235 and *Salmonella typhosa* LPS. The BXH strains were also tested for in vivo adjuvant response to LPS. The immune response to the *o*-polysaccharide antigens on LPS was used to determine whether the lipid A region of LPS possessed adjuvant properties. The magnitude of this immune response to LPS in C3H/HeJ mice and in other strains of mice is greatest at high antigenic concentrations of LPS (14). Using 100 μg LPS, a small antibody response to LPS develops in C3H/HeJ mice, but this is generally much lower when compared with other strains of mice (14). Although some variability in antibody titer is observed for the BXH strains after seven days, mice from BXH-5, 11, 14, and 19 strains all developed antibody titers in excess of 1:320 to LPS, whereas mice from the remaining BXH strains had antibody titers no greater than 1:80 (Table 1). We consider antibody titers of 1:80 to be low (14). The BXH-10 mice showed small immune and mitogenic responses to LPS (Table 1) and are now considered LPS responders (14). The nine remaining BXH strains did not respond to LPS either in the mitogenic or adjuvant assays.

The data presented in Table 2 summarize the *H-2, Ig-1* and *Mup-1* genotypes of the 14 BXH RI strains. Four of the 14 BXH RI strains exhibit recombination of the LPS response character with *H-2* (BXH-4, 5, 8, and 9), and *Ig-1* (BXH-3, 8, 14, and 19). These results are consistent with independent segregation, in agreement with the backcross linkage analyses previously described (14, 19). Thirteen of the 14 BXH RI strains exhibited concordance inheritance between the *Mup-1* and LPS responsiveness. Only a single strain (BXH-18) possesses a recombinant genotype with respect to LPS responsiveness and *Mup-1*. This degree of concordance is formally significant ($p < 0.01$), suggesting linkage of the two characters.

We have utilized a backcross analysis to establish that these loci are genetically linked (25). The data from backcross (C3H/HeJ × C57BL/6J) F_1 × C3H/HeJ mice are summarized in Table 3. Only seven presumed recombinants between *Mup-1* and *Lps* were found among the 70 backcross mice, indicating linkage. Occasionally the identification of the electrophoretic variants of *Mup-1* is extremely difficult (Figure 1). Of the putative recombinant mice, two were difficult to type definitively for *Mup-1*. The recombinant frequency for *Mup-1* and *Lps* can be calculated from the backcross. Using $r = 7/70$ (24), a value of 0.10 ± 0.04 is obtained. However, since two recombinants detected in the backcross were difficult to type, we may set $r = 5/70$, yielding 0.07 as the lower limit of the recombinant frequency. This is not significantly ($p > 0.05$) higher than the data obtained from the RI strains, where 1/14 strains showed a recombinant phenotype (24). To calculate the recombination frequency for the re-

Table 2. Summary of *H-2*, *Ig-1* allotype and major urinary protein (*Mup-1*) characteristics of recombinant inbred C57BL/6J × C3H/HeJ lines[a]

RI Line	*H-2*	*Ig-1*	*Mup-1*	LPS (responder/ nonresponder
BXH-2	*k*	*a*	*a*	NR
BXH-3	*k*	*b*	*a*	NR
BXH-4	*b*	*a*	*a*	NR
BXH-5	*k*	*b*	*b*	R
BXH-6	*k*	*a*	*a*	NR
BXH-7	*k*	*a*	*a*	NR
BXH-8	*b*	*b*	*a*	NR
BXH-9	*b*	*a*	*a*	NR
BXH-10	*b*	*b*	*b*	R
BXH-11	*b*	*b*	*b*	R
BXH-12	*k*	*a*	*a*	NR
BXH-14	*k*	*b*	*b*	R
BXH-18	*k*	*a*	*b*	NR
BXH-19	*b*	*a*	*b*	R

[a] Parental types: C57BL/6J $H\text{-}2^b$, $Ig\text{-}1^b$, $Mup\text{-}1^b$. C3H/HeJ $H\text{-}2^k$, $Ig\text{-}1^a$, $Mup\text{-}1^a$.

combinant inbred lines it is necessary to employ the Haldane equation (26), in which the probability of fixing a recombinant genotype (r) is $4r/(1 + 6r)$ where r is the probability of recombination in a single meiosis. Using $r = 1/14$, a value of 0.02 ± 0.02 is obtained for the recombination frequency (24). A weighted average of the backcross and RI estimated recombination frequencies is 0.04 ± 0.02.

The *Mup-1* locus is linked to the brown coat color (*b*) on mouse chromosome 4 (27). Thus the location of the defective LPS response gene in C3H/HeJ mice is on chromosome 4. We propose the locus

Table 3. Summary of backcross linkage analysis[a]

Genotype	Number of high responder mice (Lps^n/Lps^d)	Number of low responder mice (Lps^d/Lps^d)
$Mup\text{-}1^a/Mup\text{-}1^a$	3	33
$Mup\text{-}1^a/Mup\text{-}1^b$	30	4

[a](C3H/HeJ × C57BL/6J)F_1, mice were backcrossed to C3H/HeJ and the progeny were tested for electrophoretic variants of the major urinary protein. C3H/HeJ is $Mup\text{-}1^a$ and C57BL/6J is $Mup\text{-}1^b$. Backcross mice were either homozygous ($Mup\text{-}1^a/Mup\text{-}1^a$) or heterozygous ($Mup\text{-}1^a/Mup\text{-}1^b$) at this locus. We use the locus symbol *Lps*, with the mutant allele of C3H/HeJ designated as Lps^d and the normal allele of other strains designated Lps^n. These data are from Reference 19.

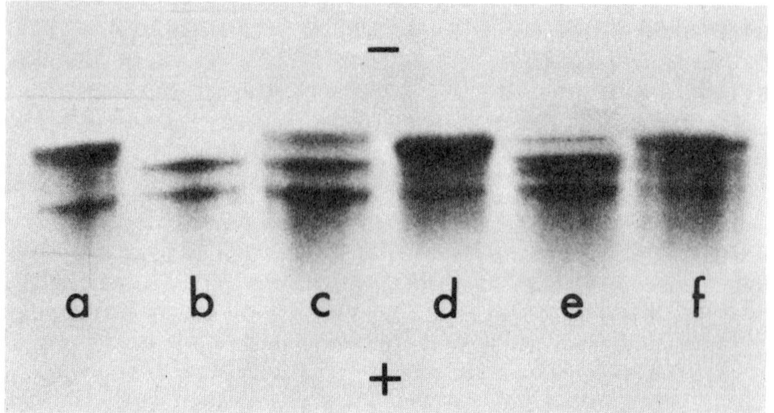

Figure 1. Electrophoretic separation of the major urinary protein. Samples were prepared and subjected to electrophoresis as described under Methods. Wells a,b, and c contained urine samples from female C3H/HeJ, C57BL/6J and F_1 (C3H/HeJ × C57BL/6J) mice, respectively. Wells d,e, and f contained urine samples from male C3H/HeJ, C57BL/6J, and F_1 (C3H/HeJ × C57BL/6J) mice, respectively. The direction of migration of the proteins is toward the anode.

symbol *Lps* with the mutant allele of C3H/HeJ designated Lps^d (defective response) and the normal allele of other strains designated Lps^n (normal response) (26).

We have used the locus for *Ps* (polysyndactyly) (28) as a marker to determine the location of this LPS response gene on chromosome 4. F_1 hybrid mice from C57BL/6J-By-Ps × C3H/HeJ parents were backcrossed to C3H/HeJ (25). The segregation of phenotypes for *Mup-1*, *Lps*, and *Ps* markers in the backcross progeny are shown in Table 4, placing *Lps* between *Mup-1* and *Ps*. Previous studies have mapped the *Mup-1* and *Ps* genes on the centromeric and distal sides of *b*, respectively (28, 29). The recombination frequency between *Mup-1* and *b* has been reported as 0.04 ± 0.02 (29), and between *b* and *Ps* as 0.08 (28). Our estimate of the distance between *Mup*-1 and *Lps* is a value of 0.04 ± 0.02 (25). This would place *Lps* very close to the *b* locus (Figure 2).

THE EXPRESSION OF THE LPS LOCUS

Our studies on the effect of LPS on B lymphocytes have led to the general reactions to LPS in mice that constitute the endotoxic response. In addition to its resistance to the mitogenic, polyclonal, and adjuvant effects of LPS (14, 15, 23), C3H/HeJ mice also show resistance to a variety of nonimmunologic effects such as toxicity (22), non-specific resistance to bacterial infections (30), macrophage activation (30, 31), and increases in levels of various serum components such as colony stimulating factor (CSF) (32) and the acute phase serum amyloid protein (SAA) (33). The major problem in attempting to correlate the genetic control of LPS responses in B lymphocytes to those involving the interaction of LPS with other cell types is the difficulty in performing more than one type of LPS response assay in individual animals. We have examined three nonimmunologic responses induced by LPS, utilizing 12 of the recombinant inbred strains of mice derived from C3H/HeJ and C57BL/6J parental strains, and a backcross linkage analysis. The first response is a hypothermia induced by LPS when mice are placed in an environment below 25°C, which is in contrast to a hyperthermia when the temperature is greater than 30°C (34). The mechanisms regulating these fluctuations in body temperature are not known. The second response is the production of a colony stimulating factor (CSF) which stimulates granulocyte and macrophage colony formation by mouse bone marrow cells in agar cultures (35). LPS does not elevate serum CSF levels in C3H/HeJ mice as in other strains of mice (32). The third LPS response examined is the production of a serum precursor SAA (36) of the secondary amyloid protein AA (33).

All recombinant inbred strains that express *Lps*^{d} (BXH-2, 3, 4, 6, 7, 8, 9, 12) do not respond to an injection of LPS by a hypothermia, or by increases in serum levels of either CSF or SAA (37). The *Lps*^{n}

Table 4. Segregation of *Mup-1*, *Lps*, and *Ps* markers in backcross (C3H/HeJ × C57BL/6J-By-Ps)F_1 × C3H/HeJ mice

Genotypes[a]	*Mup-1*	*Lps*	*Ps*	*N* Mice
C3H/HeJ	*a*	*d*	+	15
Heterozygote	*b*	*n*	*Ps*	10
Recombinants	*a*	*d*	*Ps*	2
	b	*n*	+	2
	a	*n*	*Ps*	1
	b	*d*	+	1
	a	*n*	+	0
	b	*d*	*Ps*	0

[a]The genotypes for *Mup-1*, *Lps*, and *Ps* markers, respectively are: C3H/HeJ, *a d* +/*a d* +; C57BL/6J-By Ps, *b n Ps*/*b n Ps*; F_1, *a d* +/*b n Ps*. Backcross mice were first typed for *Mup-1* and *Ps*, and then spleen cultures were assayed for mitogenic responsiveness to LPS. These data are summarized from Reference 19.

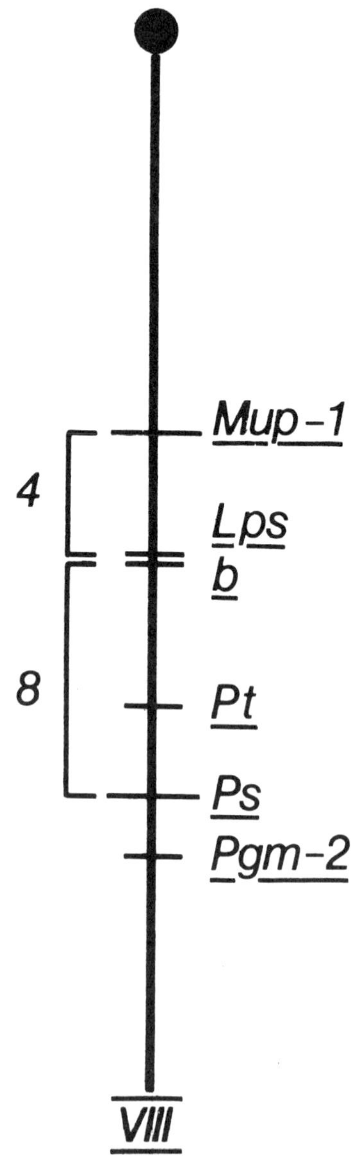

Figure 2. The location of the *Lps* locus on chromosome 4 of the mouse.

strains (BXH-5, 11, 14, and 19) all show hypothermal and CSF responses to LPS. In three of these RI strains, BXH-5, 14, and 19, LPS induced large increases in SAA levels. However, one *Lps*^{n} strain, BXH-11, shows only a slight SAA response to LPS, which may reflect the involvement of a genetic locus in the synthesis of SAA that is distinct from *Lps*. This locus is apparently altered in BXH-11 mice (37).

A backcross linkage analysis has been used to examine the involvement of *Lps* in those same three LPS-induced responses. F_1 (C3H/HeJ × C57BL/6J mice were backcrossed to C3H/HeJ mice and the individual progeny were examined in a number of ways. First, backcross progeny with a homozygous *Mup-1*^{a}/*Mup-1*^{a} phenotype did not show either hypothermal responses to an injection of LPS or increases in serum CSF or SAA. However, backcross progeny with heterozygous *Mup-1*^{a}/*Mup-1*^{b} phenotypes were LPS responders in each of the three assays (37). Several recombinant phenotypes were observed, as would be expected upon our estimation of the recombination frequency between *Mup-1* and *Lps*. Thus the expression of *Mup-1* seems to be an accurate genetic marker for the linked locus, *Lps*. To verify this, we have kept mice for 3 weeks after a single injection of LPS and subsequent analysis of CSF or SAA levels, and then assayed mitogenic responses to LPS in spleen cultures. We know the mitogenic response to be a direct assay of the expression of the *Lps* locus (14). With the exception of several presumed recombinant backcross progeny, all *Mup-1*^{a}/*Mup-1*^{a} mice were *Lps*^{d}, whereas *Mup-1*^{a}/*Mup-1*^{b} mice were *Lps*^{n}. Therefore, there was concordance in the expression of *Mup-1* and *Lps* in each LPS response assay (37).

THE *Lps* LOCUS AND ENDOTOXIC REACTIONS

To initiate a specific biochemical response, LPS must interact with a cellular structure that may be termed a LPS receptor. Our genetic studies have shown that the expression of the *Lps* locus is required for the initiation of LPS responses in many different cell types. The simplest interpretation of these findings is that *Lps* is involved in the production of a LPS receptor, or the activation of an enzyme through which the LPS receptor exerts its response. A LPS receptor has been isolated from human erythrocytes (38). Whether this structure is also found in various murine cell types and is the cellular structure through which LPS initiates the various cellular responses awaits study. The primary effect of the lipid A moiety of LPS in eliciting the immunologic and nonimmunologic responses that constitute the endotoxic reactions is the specific activation of the product of the *Lps* locus. This seems to be an initiation event common to the cell types that respond to LPS. Many cellular effector systems may be activated as a consequence of this primary reaction. However there may be many reactions, such as the initiation of acute inflammatory responses and intravascular coagulation, that constitute secondary responses to LPS, since these reactions may be mediated by a wide range of chemical mediators that are subsequently released by LPS-activated cells.

The importance of understanding the *Lps* locus lies not only in its use as a tool to dissect the control of the immune system, but in its use in a protective capacity against endotoxic effects during infection with Gram-negative bacteria. Humoral antibody directed against the *o*-antigens has little protective activity against the endotoxic effects of LPS. However, knowledge of the *Lps* locus may lead to the development of metabolic inhibitors to block the primary response of cells to lipid A, thereby preventing the initiation of endotoxic reactions. Our genetic studies indicate that this approach is feasible since there is a common cellular mechanism involved in the initiation of the many diverse responses to lipid A.

LITERATURE CITED

1. Landy, M., and Baker, P. J. 1966. Cytodynamics of the distinctive immune response produced in regional lymph nodes by Salmonella somatic polysaccharide. J. Immunol. 97:670.
2. Rudbach, J. A. 1971. Molecular immunogenicity of bacterial lipopolysaccharide antigens: Establishing a quantitative system. J. Immunol. 106:993.
3. Skidmore, B., Chiller, J., Morrison, D., and Weigle, W. 1975. Immunologic properties of bacterial lipopolysaccharide (LPS): Correlation between the mitogenic, adjuvant, and immunogenic activities. J. Immunol. 114:770.
4. Clamen, H. 1963. Tolerance to a protein antigen in adult mice and the effect of nonspecific factors. J. Immunol. 91:833.
5. Luderitz, O., Westphal, O., Staub, A. M., and Nikaido, H. 1971. Isolation and chemical and immunological characterization of bacterial lipopolysaccharides. In G. Weinbaum, S. Kadis, and S. J. Ajl (eds.), Microbiol Toxins, p. 145. Academic Press, New York.
6. Anderson, J., Moller, G., and Sjoberg, O. 1972. Selective induction of DNA synthesis in T and B lymphocytes. Cell. Immunol. 4:381.
7. Watson, J., Trenkner, E., and Cohn, M. 1973. The use of bacterial lipopolysaccharides to show that two signals are required for the induction of antibody synthesis. J. Exp. Med. 138:699.
8. Von Eschen, K. B., and Rudbach, J. A. 1974. Immunological responses of mice to native protoplasmic polysaccharide and lipopolysaccharide. J. Exp. Med. 140:1604.

9. Louis, J., Chiller, J., and Weigle, W. 1973. The ability of bacterial lipopolysaccharide to modulate the induction of unresponsiveness to a state of immunity. J. Exp. Med. 138:1481-1485.
10. Sidman, C. L., and Unanue, E. R. 1975. Development of B lymphocytes. I. Cell populations and a critical event during ontogeny. J. Immunol. 114:1730.
11. Hammerling, U., Chin, A. F., and Abbott, J. 1976. Ontogeny of murine B lymphocytes: Sequence of B cell differentiation from surface-immunoglobulin-negative precursors to plasma cells. Proc. Nat. Acad. Sci. USA 73:2008.
12. Hammerling, U., Chin, A. F., Abbott, J., and Scheid, M. P. 1975. The ontogeny of murine B lymphocytes. I. Induction of phenotypic conversion of Ia^- to Ia^+ lymphocytes. J. Immunol. 115:1425.
13. Sultzer, B. M., and Nilsson, B. S. 1972. PPD tuberculin—a B cell mitogen. Nat. New Biol. 240:199.
14. Watson, J., and Riblet, R. 1974. Genetic control of responses to bacterial lipopolysaccharides in mice. I. Evidence for a single gene that influences mitogenic and immunogenic responses to lipopolysaccharides. J. Exp. Med. 140:1147.
15. Watson, J., and Riblet, R. 1975. Genetic control of responses to bacterial lipopolysaccharides in mice. II. A gene that influences a membrane component involved in the activation of bone marrow-derived lymphocytes by lipopolysaccharides. J. Immunol. 114:1462.
16. Watson, J., Riblet, R., Cohn, M., Skidmore, B., Chiller, J., and Weigle, W. 1976. The mitogenic activity of bacterial lipopolysaccharide as a probe for T cell function. In J. J. Oppenheim and D. L. Rosenstreich (eds.), Mitogens in Immunobiology, pp. 245-260. Academic Press, New York.
17. Komuro, K., and Boyse, E. A. 1973. Induction of T lymphocytes from precursor cells *in vitro* by a product of the thymus. J. Exp. Med. 138:479.
18. Watson, J. 1977. Differentiation of B lymphocytes in C3H/HeJ mice. The induction of Ia antigens by lipopolysaccharide. J. Immunol. 118:1103.
19. Kelly, K., and Watson, J. 1977. The inheritance of a defective lipopolysaccharide response to locus in C3H/HeJ mice. Immunogenetics 4:26.
20. Coutinho, A., Moller, G., and Gronowicz, E. 1975. Genetic control of B cell responses. IV. Inheritance of the unresponsiveness to lipopolysaccharides. J. Exp. Med. 142:253-258.
21. Glode, L. M., and Rosenstreich, D. L. 1976. Genetic control of B cell activation by bacterial lipopolysaccharide B mediated by multiple distinct genes or alleles. J. Immunol. 117:2061.
22. Sultzer, B. M. 1976. Genetic analysis of lymphocyte activation by lipopolysaccharide endotoxin. Infect. Immun. 13:1579-1584.
23. Skidmore, B. J., Chiller, J. M., Weigle, W. O., Riblet, R., and Watson, J. 1976. Immunologic properties of bacterial lipopolysaccharide (LPS). III. Genetic linkage between the in vitro mitogenic and in vivo adjuvant properties of LPS. J. Exp. Med. 143:143.
24. Watson, J., Riblet, R., and Taylor, B. A. 1977. The response of recombinant inbred strains of mice to bacterial lipopolysaccharide. J. Immunol. In press.
25. Watson, J., Kelly, K., Largen, M., and Taylor, B. A. 1978. The genetic

mapping of a defective LPS response gene in C3H/HeJ mice. J. Exp. Med. J. Immunol. 120:422.
26. Haldane, J. B. S., and Waddington, C. H. 1931. Inbreeding and linkage. Genetics 16:357.
27. Hudson, D. M., Finlayson, J. S., and Potter, M. 1967. Linkage of one component of the major urinary protein complex of mice to the brown coat color locus. Genet. Res. 10:195.
28. Johnson, D. R. 1969. Polysyndactyly, a new mutant gene in the mouse. J. Embryol. Expt. Morphol. 21:285.
29. Finlayson, J. S., Hudson, D. M., and Armstrong, B. L. 1969. Location of *Mup-a* of mouse in linkage group VIII. Genet. Res. 14:329.
30. Chedid, L., Parant, M., Damais, C., Parant, F., Juy, D., and Galelli, A. 1976. Failure of endotoxin to increase nonspecific resistance to infection of lipopolysaccharide low responder mice. Infect. Immun. 13:722.
31. Sultzer, B. M. 1968. Genetic control of leukocyte responses to endotoxin. Nature 219:1253.
32. Apte, R. N., and Pluznik, D. H. 1968. Genetic control of lipopolysaccharide induced generation of serum colony stimulating factor and proliferation of spenic granulocyte/macrophage precursor cells. J. Cell. Physiol. 89:313.
33. McAdam, K. P. W. J., and Sipe, J. D. 1976. Murine model for human secondary amyleidosis: Genetic variability of the acute phase serum protein SAA response to endotoxins and casein. J. Exp. Med. 144:1121.
34. Berry, L. J. 1966. Effect of environmental temperature on lethality of endotoxin and its effect on body temperature in mice. Fed. Proc. 25:1264.
35. Metcalf, D. 1971. Acute antigen-induced elevation of serum colony stimulating factor (CSF) levels. Immunology 21:427.
36. Rosenthal, J., Franklin, E. C., Frangeine, B., and Greenspan, J. 1976. Isolation and partial characterization of SAA and amyloid related protein from human serum. J. Immunology 116:1415.
37. Watson, J., Largen, M., and McAdam, K. P. W. J. 1978. Genetic control of endotoxic responses in mice. J. Exp. Med. 147:39.
38. Springer, G. F., Adye, J. C., Bezhoravainy, and Hirgenson, B. 1974. Properties and activity of the lipopolysaccharide receptor from human erythrocytes. Biochemistry 13:1379-1389.

Infection, Immunity, and Genetics
Edited by Herman Friedman, T. Juhani Linna, and James E. Prier

GENETIC CONTROL OF DEFENSES AGAINST INTRACELLULAR PATHOGENS

Stephen I. Vas

It has been noticed since the earliest days of history that the susceptibility of populations to infections during epidemics varies. This was considered to be the result of environmental factors, closeness of contact, or the infectious dose. Although all these play important roles in the transmission and incidence of infectious diseases, it has been recognized only recently that genetic factors may also play a role in individual susceptibility. It has been known to animal breeders for many centuries that selective breeding may increase the resistance of various animals to infections. They have used this knowledge to their advantage, although the underlying mechanisms were unknown to them. A number of genetically controlled immunological deficiencies in humans have been recognized lately, such as congenital complement deficiencies, phagocytic disfunction syndromes, and congenital defects of the B or T cell systems, and it was found that in a number of cases these defects may result in increased susceptibility. The increase of susceptibility in these cases is not specific; it extends to a wide range of infectious agents. More recently, developing knowledge on the immune regulatory gene offered explanations to the control of antibody production against certain antigens. The infectious process, the defense mechanisms, or the ultimate lethal outcome of certain bacterial infections are extremely complex events. Diseases caused by facultative intracellular parasites, a classic example of which is typhoid fever or brucellosis in man, are the prime examples of such infections. Our understanding of the disease process and immune mechanisms in such infections is limited at present.

Experimental model systems have been available, primarily based on animal breeders' experiences. Hill (1) in 1934 summarized the information available at that time on experiments in rodents and fowls. In 1937 Webster (2) at the Rockefeller Institute described mouse strains from their colonies which showed inherited differences against bacterial and virus infections. These strains became known as BSVS and BRVR strains and are presently still used by some investigators (7). Gowen and Calhoun (3) in 1943 summarized their experience with the susceptibility of certain inbred mouse strains to *Salmonella typhimurium* infections; Bell (4) in 1949 discussed resistance of various strains of chicks. In 1972, Robson and Vas (5) found marked differences in the susceptibility between inbred mouse strains widely used for scientific research (Figure 1). Work done by them and by Plant and Glynn (6) pointed to the genetic nature of these differences. Other groups of workers found differences in the susceptibility of inbred mice to other microorganisms (8–14). Some of this information is summarized in Table 1 from experimental work in which the purpose was to compare genetic differences of susceptibility to infections. A large amount of incidental information is available in the literature as well. No clear-cut pattern of resistance emerges from this information. Some of the mouse strains that have been tested with different pathogens show extreme susceptibility to one but considerable resistance to another organism (5, see Table 2). Therefore, it seems that susceptibility or resistance is dependent not only on the genetic makeup of the mouse strain but also on the infecting organism. Outbred mouse strains that were used primarily as control animals show a higher degree of resistance.

The nature of mouse typhoid has been well studied, but the events during disease and the nature of immunity are complex. A simplified scheme of factors interacting in concert are shown in Figure 2. Many or all of these factors are functional during the course of disease in mice interacting at various times. In a susceptible animal like C57BL/6 the LD_{50} of *S. typhimurium* is below ten organisms and certainly approaching the theoretical number of one bacterium per mouse. The infection in these animals is invariably fatal. The animal becomes septicemic, the number of *S. typhimurium* organisms in the liver and in the spleen increases daily, and the animal dies in few days. Immunization attempts in these susceptible animals, whether with a phenol-killed vaccine or with live avirulent microorganisms (5), will lead to some prolongation but will not protect from the ultimate outcome. C57BL/6 mice tolerate infections with live avirulent *S. typhimurium* quite well but succumb to the virulent organism. The nature of the virulence factor is unknown. The course of infection is different in a resistant animal like A/J. Although large infecting doses will kill

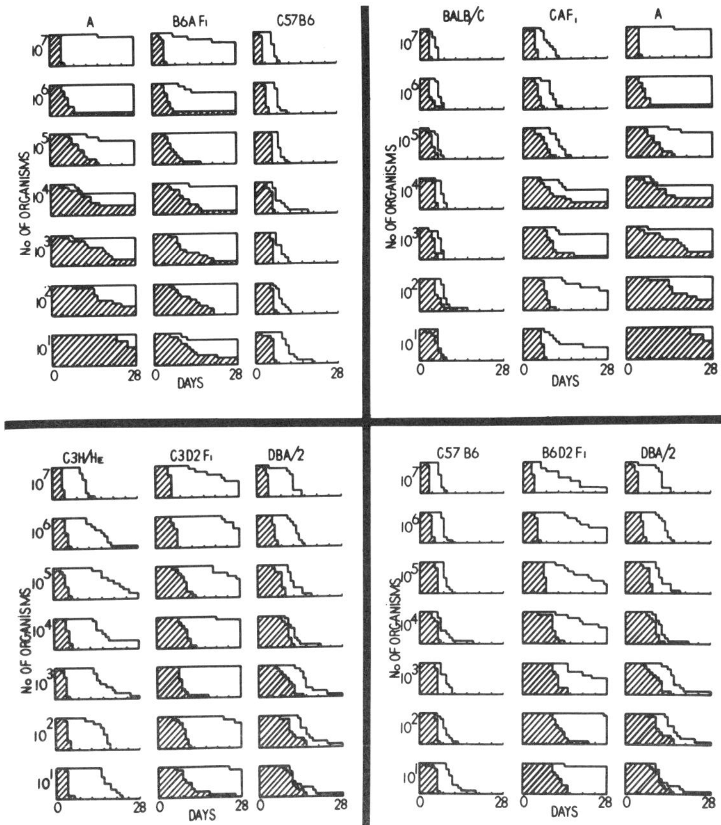

Figure 1. Survival of inbred and F1 hybrid mice infected intraperitoneally (i.p.) with various doses of *S. typhimurium*. Control mice are in shaded blocks, mice vaccinated with phenol-killed *S. typhimurium* in open blocks. Each block represents ten mice. Reprinted with permission from Robson, H. G., and Vas, S. I. 1972. J. Inf. Dis. 126:378-386. Copyright© 1972 by the University of Chicago.

control animals within a few days, similar to C57BL/6, immunized animals will show a limitation of bacterial multiplication as measured in the blood, liver, or spleen and will limit the numbers of bacteria in the animal considerably in a few days. Many or maybe all of the surviving animals remain chronic carriers of the organism. *S. typhimurium* can be shown in the tissue of these animals after many months. If the resistant mouse strain is infected with a small number of challenge organisms (10^2 or 10^3), the animal will limit the numbers of organisms in its tissues similarly to that of the vaccinated animal (5).

Table 1. Susceptibility of mice to various microbial infections

Strain	Susceptible	Resistant
Inbred		
A/J	*L. monocytogenes* (5)	*S. typhimurium* (5)
	F. tularensis (9)	*H. capsulatum* (10)
	C. neoformans (5)	
A/HeJ	*H. capsulatum* (10)	
	S. typhimurium (15)	
AKR/J		*F. tularensis* (9)
BALB/cJ	*S. typhimurium* (5)	*F. tularensis* (9)
	C. kutscheri (8)	
BRVR		*S. typhimurium* (2, 24)
BSVS	*S. typhimurium* (2, 24)	
CBA/J	*F. tularensis* (9)	*S. typhimurium* (6)
CBA/N	*S. typhimurium* (16)	
C3H/J	*F. tularensis* (9)	
	M. tuberculosis (11)	
C3H/HeJ	*S. typhimurium* (5)	*S. typhimurium* (6)
	C. kutscheri (8)	
C3H/CWB	*C. kutscheri* (8)	
C3H/CSW	*C. kutscheri* (8)	
C57BL		*C. kutscheri* (8)
C57BL/6J	*S. typhimurium* (5)	*L. monocytogenes* (5)
	M. tuberculosis (14)	*C. kutscheri* (8)
		F. tularensis (9)
		C. neoformans (5)
C57BL/10A	*C. kutscheri* (8)	
C57BL/10Br	*C. kutscheri* (8)	
C57BL/10D2	*C. kutscheri* (8)	
C57BL/10Sn	*H. capsulatum* (10)	
DBA/1	*M. tuberculosis* (11)	
DBA/2J	*C. kutscheri* (8)	*S. typhimurium* (5)
Strong 1	*M. tuberculosis* (11)	
Strong A		*M. tuberculosis* (11)
Strong C		*M. tuberculosis* (11)
SWR/J		*F. tularensis* (9)
		H. capsulatum (10)
SJL/J		*F. tularensis* (9)
		H. capsulatum (10)
Outbred		
SW	*C. kutscheri* (8)	*S. typhimurium* (5)
		M. tuberculosis (11)
CF-1		*M. tuberculosis* (13)
CD-1		*F. tularensis* (9)

Mice cannot be protected passively with sera from immunized animals containing high antibody levels but the resistant strain can be protected by passive transfer of immune spleen cells (9). It is generally believed that the limiting of the multiplication of organisms in the animal is done through the macrophage activation mechanism. Anti-

Table 2. Response of C57BL/6J and A/J mice to infections

Challenge	C57BL/6J	A/J
S. typhimurium challenge	Susceptible	Resistant
Immunization with *S. typhimurium* phenol vaccine	Not protected	Protected
L. monocytogenes challenge	Resistant	Susceptible
Immunization with *L. monocytogenes* phenol vaccine	Protected	Not protected
Sensitivity to endotoxin	Resistant	Sensitive
S. typhimurium antigen:		
Serum antibody response	Good	Good
Plaque forming cells (Jerne)	Good	Good
Challenge with:		
D. pneumoniae	Susceptible	Susceptible
Toxoplasma gondii	Susceptible	Susceptible
C. neoformans	Resistant	Susceptible

bodies may play a role in the initial period of infection by slowing down the rate of bacterial multiplication until macrophage activation is achieved, usually by the fourth or fifth day post challenge. It is believed that development of delayed type hypersensitivity against *Salmonella* antigens is important in the immune defense of mice, as summarized by Collins and Makaness (17).

There has been no doubt from the beginning concerning the genetic nature of control mechanisms of resistance, but the finer

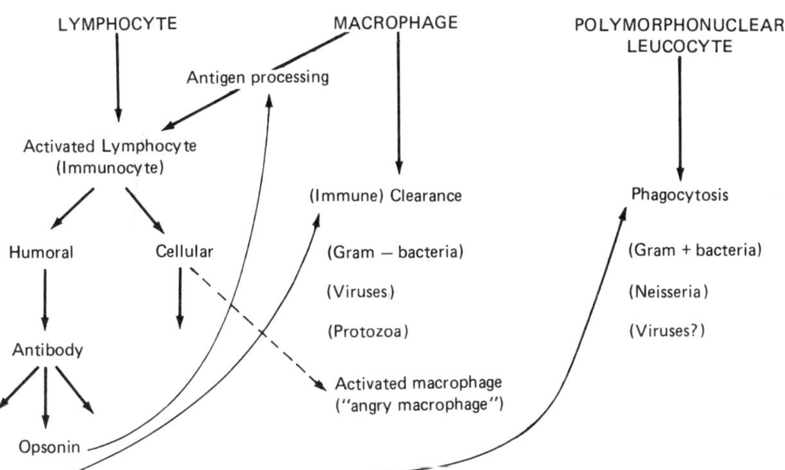

Figure 2. Cellular mechanisms of antibacterial immunity.

analysis of these mechanisms has eluded the researchers as yet. Plant and Glynn (6) have presented evidence recently on some genetic analysis of susceptible and resistant mouse strains. F_1, F_2, and parental back-cross generations from CBA and BALB/c mice showed, according to their interpretation, a simple Mendelian dominant control of resistance. It was not linked to *H-2* genes, although according to the authors it may be operational through control of delayed hypersensitivity to some unknown antigen in *Salmonella*. Similar genetic analysis was performed by Hirst and Wallace (8) on inbred mice using *Corynebacterium kutscheri* as an infecting agent. They believed that this bacterium is eliminated by microbicidal action of the mononuclear phagocytes in the liver. They concluded that the resistance of the mice against *C. kutscheri* measured by challenge was under polygenic control and was inherited continuously. The efficiency of clearance of the organism by liver macrophages was inherited discontinuously and was controlled by a single dominant autosomal gene. They gave the recessive allele the symbol *ack* (anti-*C. kutscheri*). They found no association between this locus and the immune-response genes.

There are inherent difficulties in doing genetic research with these model systems. The difficulty lies in the destructive nature of the test; the animals have to be challenged in order to assess their resistance. If the survivors are bred further, there is a great risk of selecting latent carriers and therefore of falsifying any breeding experiment. This difficulty has been pointed out by Hills' early classic review (1). No independent markers are available at present. Pierce, Dubos, and Middlebrook (11) considered coat color to be an available marker in infections with *Mycobacterium tuberculosis*. Such marker is not available in the *Salmonella* or *Corynebacterium* model system. It is further possible that separate marker will be needed for each infecting organism since evidence (5) shows that resistance against different microorganisms is inherited separately. Similarly there is no indication of the gene product. Therefore, relatively inexpensive and simple in vitro tests are not available for the genetic analysis.

The mechanisms through which genetic control may act are numerous (Figure 2). No single defect in any of the mechanisms identified up to this point is sufficient to explain the existing large differences in susceptibility. In addition, where such defects have been identified they are operational, against only one pathogenic organism. Hirst and Wallace (8) demonstrated a decrease in the efficiency of microbicidal action of the mononuclear phagocytes in the liver in C57BL/6 animals, which are sensitive compared to A/J animals which are relatively resistant to *C. kutscheri* infections. Although the same relationship is true for *S. typhimurium* infections, as shown by Robson and Vas (5) (C57BL/6 sensitive, A/J resistant), the reverse is true in these animals for *Listeria monocytogenes* infections. Therefore, one would have to assume that there are different mechanisms for intracellular bacterial killing in the macrophage. This is not impossible, but one would have to then search for the recognition and control mechanisms that make this differentiation between microorganisms possible. Although this may be antigen-directed and T cell-mediated, the direct evidence for this is lacking at present. A summary of possible sites of control is given in Table 3.

Auzins and Rowley (18) show evidence that BALB/c mice fail to respond to antigen 5 and have less response to antigen 1, 4, and 12 of the somatic antigen of *S. typhimurium* than Swiss white mice.

The work of Scher, O'Brien, and Fornal (16) on CBA/N mice, a strain that has an X-linked abnormality in the development of B lymphocytes, shows some promise. This strain is susceptible to *S. typhimurium* infections. Crossbreeding with DBA/2 (resistant) mice indicates the involvement of the X chromosome in resistance. Robson and Vas (5) data, on the other hand, showed that sufficient amounts of antibodies are produced by immunized C57BL/6 mice, but their survival was only marginally prolonged (Figure 1). Similarly, transfer of passive antibody in susceptible mice does not protect significantly. Further experiments on CBA/N mice, especially on mice actively immunized, will shed further light on this X-linked defect.

Biozzi (19) presented experiments that raised some speculative possibilities. His "high responder" mice are more sensitive to *S.*

Table 3. Possible sites of genetic control of resistance of mice against bacterial infections

B cell	Antigen sensitivity (18)
	Facilitating antibodies (19)
	Mitogenic response (20)
	Antibody affinity (21)
T cell	Delayed hypersensitivity (6)
	T - MP interaction (22, 23)
Macrophage	Phagocytosis (24)
	Activation (25)
	RES clearance (26)
	Number
	Recruitment

typhimurium or *Yersinia pestis* infections than the "low responder" strain. Immunization will protect the low responder, but it will cause only some increase in survival time in the high responder mice. He speculated on the possibility of the action of facilitating antibodies that may protect the challenge organisms against initial killing macrophages.

Von Jeney, Gunther, and Jann (20) have found that the response to the mitogenic effect of lipopolysaccharide of Gram-negative bacteria correlated well with the resistance of inbred mouse strains to intraperitoneal *S. typhimurium* challenge.

An interesting possibility is raised if one assumes different efficiencies of antibody produced by inbred mice. Alpers and colleagues (21) found that different inbred mouse strains immunized with an antigen (HSA) will produce antibody with different affinity; high affinity antibody will eliminate antigen from the circulation more rapidly than low affinity antibodies. The genetic control of this difference has not been studied. Authors also have not assumed differences in macrophage functions in these animals; these may also account for the differences in elimination of the antigen. Whether there are antibodies with differing affinity against complex bacterial antigens has not been studied in a similar system.

The explanations that may lie in differences of the thymus-derived cells of inbred mouse strains are equally difficult to evaluate. Plant and Glynn (6) have found in their genetic analysis that differences in resistance correlate to differences in delayed hypersensitivity of mouse strains to *Salmonella* antigens. The role of delayed hypersensitivity has been well studied (17) for *S. typhimurium* immunity in mice.

The role of T cells and T cell-macrophage interaction is an important link in the defense mechanism of mice against intracellular infections. Although experimental evidence is lacking in this field, interesting possibilities are raised by experiments performed on *nude* mice with *L. monocytogenes* infections (22, 23). *Nude* mice (*nu/nu*), which are congenitally hypothymic, will remain chronically infected with *L. monocytogenes*, whereas euthymic mice will show an initial rise in numbers of *Listeria* organisms in the challenged animals, then a decrease and finally elimination of the challenge organism. Moreover, the LD_{50} is much higher for the *nude* mice than for normal control mice. If chronically infected *nude* mice are given spleen cells from immune control mice, *Listeria* is eliminated from the *nude* mice within one week. These results would imply that a defect in the T cells alone would not give adequate explanation for differences in susceptibility, although the important role of T cells in controlling macrophage functions are underlined by these experiments.

The role of the macrophage in the resistance against intracellular infections is central. The possibility that the major defect is in the microbicidal action of macrophages has already been alluded to by Hirst and Wallace (8). Maier and Oels (24) have shown that although BRVR and BSVS mice are equally efficient in phagocytosing *S. typhimurium* in vitro, they show differences in their abilities to kill the intracellular organisms. Studies in the authors' laboratory (unpublished observations) could not show differences in phagocytic or killing ability between C57BL/6 and A/J mice in in vitro studies. No such differences were evident between normal peritoneal macrophages, peritoneal macrophages from immune animals or using normal peritoneal macrophages, and opsonized *S. typhimurium* in an in vitro system. The technical difficulties of studying phagocytosis and intracellular killing in peritoneal macrophages in vitro are numerous, and precaution is required to evaluate such experiments.

The ability of activated macrophages to show increased killing ability against various microorganisms has been well established. Interesting possibilities have been raised by the experiments of Medina, Vas, and Robson (25). They showed that C57BL/6, A/J mice can be protected against challenge by a number of agents, both specific and non-specific, known to activate macrophages. The protection is non-specific as far as the activating agent is concerned but can provide protection only against the challenge organisms to which the animal possesses genetically determined resistance. The authors concluded that the control may lie in the macrophage activating mechanism, possibly acting through T cell interaction.

Passwell, Steward, and Soothill (26) showed that there are differences in inbred mouse strains in reticuloendothelial clearance as measured by carbon clearance. They connected these observations with their earlier work on differences in antibody affinity (21) and discussed the possibility of the role of macrophages in antigen processing. Biozzi et al. (27) pointed out differences in antigen processing in their high responder and low responder mice. These differences were contrary to the susceptibility of their mouse strains to bacterial challenge. In our laboratory we have found C57BL/6 mice to be consistently more active in RES clearance than A/J, although C57BL/6 is much more susceptible to *S. typhimurium* challenge. Reticuloendothelial functions were measured by the elimination of ^{32}P-labeled *S. typhimurium* killed by mild phenol treatment (Table 4). It is un-

Table 4. RES clearance of inbred mouse strain using phenol-killed ^{32}P-labeled *S. typhimurium* I.V.

Mouse strain	Vaccine	Opsonizing serum	Phagocytic index ($K \pm$ S.E.)	p value
A/J			0.0155 ± 0.0007	< 0.02
C57BL/6J			0.0178 ± 0.0003	
A/J	Killed		0.0392 ± 0.0024	< 0.02
C57BL/6J	Killed		0.0458 ± 0.0015	
A/J	Live avirulent		0.0515 ± 0.0029	< 0.05
C57BL/6J	Live avirulent		0.0578 ± 0.0024	
A/J		A/J	0.0314 ± 0.0056	< 0.01
C57BL/6J		C57BL/6J	0.0516 ± 0.0015	
A/J		C57BL/6J	0.0456 ± 0.0037	< 0.01
C57BL/6J		A/J	0.0672 ± 0.0060	

likely, therefore, that RES clearance alone could be responsible for the differences in susceptibility.

The macrophage system of experimental animals is not well understood and only in recent years has research started to investigate the availability of macrophages, their numbers in the intact host, and the recruitment of macrophages to various areas of the body. The possibility that animals possess different populations of macrophages that are recruited on different stimuli cannot be excluded. Although the technical difficulties in designing the experiments to study these functions are great, such research will be forthcoming. These experiments will contribute to our understanding of genetic control of susceptibility and defense against bacterial infections.

LITERATURE CITED

1. Hill, A. B. 1934. The inheritance of resistance to bacterial infections in animal species. Spec. Rep. Ser. Med. Res. Conc. No. 196. H. M. Stationary Office, London.
2. Webster, L. T. 1937. Inheritance of resistance of mice to enteric bacterial and neurotropic virus infections. J. Exp. Med. 65:113–138.
3. Gowen, J. W., and Calhoun, M. L. 1943. Factors affecting genetic resistance of mice to mouse typhoid. J. Inf. Dis. 73:40–56.
4. Bell, A. E. 1949. Physiological factors associated with genetic resistance to fowl typhoid. J. Inf. Dis. 85:154–169.
5. Robson, H. G., and Vas, S. I. 1972. Resistance of inbred mice to *Salmonella typhimurium*. J. Inf. Dis. 126:378–386.

6. Plant, J., and Glynn, A. A. 1976. Genetics of resistance to infection with *Salmonella typhimurium* in mice. J. Inf. Dis. 133:72-78.
7. Groschel, D., Pass, C. N. S., and Rosenberg, B. S. 1970. Inherited resistance and mouse typhoid. J. Reticuloendoth. Soc. 7:484-499.
8. Hirst, R. G., and Wallace, M. E. 1976. Inherited resistance to *Corynebacterium kutscheri* in mice. Infect. Immun. 14:475-482.
9. Eigelsbach, H. T., Hunter, D. H. Janssen, W. A., Dangerfield, H. G., and Rabinowitz, S. G. 1975. Infect. Immun. 12:999-1005.
10. Chick, E. W., and Roberts, G. D. 1974. The varying susceptibility of different genetic strains of laboratory mice to *Histoplasma capsulatum*. Mycopath. Mycol. Appl. 52:251-253.
11. Pierce, C., Dubos, R. J., and Middlebrook, G. 1947. Infection of mice with mammalian tubercle bacilli grown in Tween-albumin liquid medium. J. Exp. Med. 86:159-173.
12. Donovick, R., McKee, C. M., Jambor, W. P., and Rake, G. 1949. The use of the mouse in a standardized test for antituberculous activity of compound of natural or synthetic origin. II. The choice of mouse strain. Am. Rev. Tuberc. 60:109-120.
13. Youmans, L. P., Youmans, A. S., and Kanai, K. 1959. The difference in response of four strains of mice to immunization against tuberculous infection. Am. Rev. Resp. Dis. 80:750-756.
14. Youmans, G. P., and Youmans, A. S. 1972. Response of vaccinated and non-vaccinated syngeneic C57BL/6 mice to infection with *Mycobacterium tuberculosis*. Infect. Immun. 6:748-754.
15. Vas., S. I., Roy, R., and Robson, H. G. 1973. Endotoxin sensitivity of inbred mouse strains. Can. J. Microbiol. 19:767-769.
16. Scher, I., O'Brien, A., and Fornal, S. 1977. *Salmonella typhimurium* infection in mice: Influence of x linked gene controlling B lymphocyte function. Abstr. Annual meeting of ASM. E. 48:89.
17. Collins, F. M., and Mackaness, G. B. 1968. Delayed hypersensitivity and Arthus reactivity in relation to host resistance in *Salmonella* infected mice. J. Immunol. 101:830-845.
18. Auzins, I., and Rowley, D. 1969. The response to *Salmonella typhimurium* antigens by two different strains of mice. Immunology 17:579-585.
19. Biozzi, G. 1972. In H. O. McDevitt and M. Landy (eds.), Genetic Control of Immune Responsiveness, pp. 317-322. Academic Press, New York.
20. von Jeney, N., Gunther, E., and Jann, K. 1977. Mitogenic stimulation of murine spleen cells: Relation to susceptibility to *Salmonella* infections. Infec. Immun. 15:26-33.
21. Alpers, J. H., Steward, M. W., and Soothill, J. F. 1972. Differences in immune elimination in inbred mice: The role of low affinity antibody. Clin. Exp. Immunol. 12:121-132.
22. Emmerling, P., Finger, H., and Bockemuhl, J. 1975. *Listeria monocytogenes* infection in nude mice. Infect. Immun. 12:437-439.
23. Emmerling, P., Finger, H., and Hof, F. 1977. Cell mediated resistance to infection with *Listeria monocytogenes* in nude mice. Infect. Immun. 15:382-385.

24. Medina, A., Vas, S. I., and Robson, H. G. 1975. Effect of nonspecific stimulation on the defense mechanisms of inbred mice. J. Immunol. 114:1720-25.
25. Maier, T., and Oels, H. C. 1972. Role of macrophage in natural resistance to *Salmonellosis* in mice. Infect. Immun. 6:438-443.
26. Passwell, J. H., Steward, N. W., and Soothill, J. F. 1974. Inter-mouse strain differences in macrophage function and its relationship to antibody responses. Clin. Exp. Immunol. 17:159-167.
27. Biozzi, G., Stiffel, C., Mouton, D., and Bouthillier, Y. 1975. Selection of lines of mice with high and low antibody responses to complex immunogens. In B. Benacerraf (ed.), Immunogenetics and Immunodeficiency, pp. 179-227. University Park Press, Baltimore.

Infection, Immunity, and Genetics
Edited by Herman Friedman, T. Juhani Linna, and James E. Prier

GENETIC CONTROL OF THE "TRUE" PRIMARY IMMUNE RESPONSE TO CHOLERA SOMATIC ANTIGEN

Herman Friedman

One of the major observations in contemporary immunology during the last decade or so has been the discovery of a close relationship between genetic control by certain genes and the ability of an individual to respond immunologically to various antigens (1–5). These genes are related to the histocompatibility system, not only in experimental animals but also in man. The histocompatibility system has become a major focus of attention not only for those involved in transplantation immunology, but also for those involved in biomedicine in general, since certain histocompatibility genes seem to be associated with many diseases, including infectious diseases. Earlier studies from a number of laboratories have shown that there are marked genetic differences in the ability of various inbred strains of mice to respond to specific antigens (6–8). Furthermore, the immune response to haptens and small chemical determinants have often been found to be related to histocompatibility genes, especially in mice (1–3).

Many bacterial antigens elicit different levels of immune responses in mice, rabbits, and guinea pigs because of genetic differences. In many cases either high or low responding strains of mice examined for differences in responsiveness to bacterial antigens have been studied by cross-breeding and back-breeding experiments. The relationship of the H-2 locus in mice and the genetic control of the immune response to bacteria is not, however, clear cut. This may be attributable to the multiplicity of antigenic determinants present in or on most bacterial antigens, including highly purified capsular polysaccharides, extracellular toxins, and somatic "O" antigens.

Earlier studies in this laboratory have shown that there is a genetic difference in the antibody responsiveness of inbred strains of mice to distinctive antigenic determinants present on the cell walls of *Vibrio cholerae* bacilli (10). The immune response of mice to cholera antigens is quite distinct, since all normal mice examined to date failed to demonstrate pre-existing antibody forming cells before active immunization (15–20). After immunization mice respond to the cholera bacilli in a manner that seems to be a "true primary response." No antibody forming cells are detectable until at least 42–48 hr after immunization, regardless of route or dose of cholera antigen (19–25). A rapid "stepwise" increase in antibody forming cells then occurs, with peak numbers by days 14 to 18 (16,22). However, after secondary immunization of mice already primed with antigen 6 to 10 weeks earlier, a much more rapid appearance of antibody producing cells occurs (23,24), similar to the type of response noted for other bacterial antigens such as those derived from *Escherichia coli* and *Salmonella*, as well as other ubiquitous antigens such as xenogeneic erythrocytes. Thus it seems that a true primary immune response can be readily induced to this bacterium in mice. This response is partially dependent upon T lymphocytes, with macrophages important for processing intact bacteria (25). Thus this model system seemed to lend itself quite well to the question of whether or not a genetic control system exists for expression of the immune response to the cholera bacilli in mice and whether or not such a true primary response is related to the histocompatibility system in general.

METHODS AND PROCEDURES

Inbred strains of mice were used for these studies. Most of the animals were obtained from Jackson Memorial Laboratory, Bar Harbor, Maine. The mice were housed in groups of eight to ten in plastic mouse cages and fed Purina mouse pellets and water ad libitum. In all cases the genetic homogeneity of the mouse strains was ascertained by appropriate full thickness skin grafts. All mice accepted the skin grafts essentially indefinitely from other mice within the strain.

The antibody response to *V. cholerae* somatic antigens was determined at the level of individual immunocytes (15–18). Cholera bacilli, either the Ogawa or Inaba strains, were cultured overnight in brain heart infusion broth and washed. The bacteria were resuspended to a concentration of approximately 10^9 organisms per ml. For immuni-

:ation the bacteria were heat killed by heating at 56°C for 30 min. Varying doses of the heat-killed bacteria were injected into mice, either by intraperitoneal or intravenous routes. At various times thereafter the true primary and secondary antibody responses were determined in vitro by vibriolytic plaque assays. For this purpose spleen cells were obtained from mice at various times after immunization and dispersed cell suspensions prepared by teasing with needles and forceps. The splenocytes were standardized by hemocytometer count after staining with trypan blue, and varying numbers in 0.1 ml medium were incubated in 1.0 ml melted sterile agar (Difco Noble) containing a 0.1 ml inoculum of a suspension of viable *vibrios*, either the homologous or heterologous strains. The agar containing bacteria and splenocytes was then poured onto previously prepared 60 mm diameter Petri plates containing a base layer of sterile agar. After 1–2 hr incubation at 37°C the plates were overlaid with approximately 3 ml of a 1:10 dilution of sterile guinea pig serum and incubated for an additional hour. The plates were washed with buffer and incubated for 3-5 hr until a "lawn" of bacteria developed on the surface of the agar, except in those areas of no growth because of the release of vibriolytic antibody by individual plaque forming cells (PFC). These were considered to be caused by 19S IgM PFC. In experiments designed to study the secondary immune response, rabbit anti-mouse immunoglobulin serum diluted 1:200 was incorporated into the agar so that facilitated PFC were detected; these were caused by 7S IgG antibody forming cells (23, 24).

The numbers of PFC to the common versus type-specific somatic antigens were differentiated by incubating splenocytes in duplicate; one set of plates was tested with the homologous strain of vibrios and the other set with the heterologous strain (15–18). When mice were immunized with the Ogawa strain their spleen cells were tested against the homologous Ogawa bacteria to determine the number of PFC to both antigens, and against the heterologous Inaba bacteria to determine PFC to only the common antigen present in both Ogawa and Inaba serotypes. By appropriate subtraction the numbers of PFC to the type-specific antigen could be calculated.

EXPERIMENTAL RESULTS

Groups of mice of different inbred strains were immunized with 50 μg of heated-killed vibrios. As is apparent in Table 1, markedly different antibody responses occurred in these animals when tested at the time of the expected peak response. BALB/c mice showed the highest

Table 1. Splenic vibriolytic antibody plaque response in different strains of mice

Mouse strain[a]	H-2 allele	*N* mice tested	Average vibriolytic PFC response[b]		
			Per spleen	Per 10^6 spleen cells	Ranking of response
A/jax	*a*	30	26,500 ± 5000	145 ± 36	5
C57BL/6J	*b*	48	83,400 ± 6100	479 ± 58	2
$B6AF_1$	*a* × *b*	47	14,300 ± 2100	64 ± 18	6
BALB/cJ	*d*	48	146,500 ± 4780	953 ± 150	1
DBA/2J	*d*	52	6,200 ± 1360	26 ± 7	8
B10D2/New Sn	*d*	37	9,700 ± 2400	53 ± 5	7
C3H/HeJ	*k*	60	2,158 ± 732	12 ± 3	9
C57BR/cdj	*k*	45	62,650 ± 2950	371 ± 65	3
NIH white[c]		110	53,100 ± 3700	318 ± 41	4

[a] Groups of indicated mouse strain immunized i.p. with 50 μg of heat-killed *V. cholerae* vaccine.
[b] Average total vibriolytic PFC response per spleen 12–14 days after immunization.
[c] Non-inbred mice.

response, both when the numbers of PFC were calculated per spleen and per million splenocytes tested. C3H/HeJ mice showed the lowest responses; these animals developed fewer than 10% of the total number of splenic PFC as compared to the BALB/c mice. Other mice developed varying numbers of PFC. Non-inbred animals (i.e., NIH white mice) showed approximately one-third the number of PFC as the BALB/c animals. It seemed noteworthy that there was no demonstrable relationship between the H-2 genotype of the mouse strains and their responsiveness. For example, two mouse strains with the "*d*" H-2 allele (i.e., DBA and B10.D2 mice) showed relatively modest responses as compared to BALB/c mice which shared the same H-2 genotype. Similarly, a mouse strain with the "*a*" H-2 allele (i.e., C57BR mice) showed approximately a 30-fold greater PFC response than did C3H/HeJ mice with the same allele. High responsiveness did not seem to be dominant, since B6AF1 mice (a hybrid of A/J and C57BL mice) showed fewer PFC per spleen than did either of the parental strains (Table 1).

The marked differences in antibody responses could have been attributable not only to differences in the magnitude of the response, but also to differences in the kinetics of antibody formation. For example, the peak number of PFC might not have occurred at the same time in all mouse strains. Thus other groups of inbred mice were injected with vibrio vaccine, and the numbers of PFC were determined at varying times thereafter for a period of up to 30 days. As is evident in Table 2, in all cases the peak numbers of PFC occurred at about the same time, i.e., when the assay was performed on the fourteenth day after immunization as compared to days 10 or 18. Those mouse strains that showed the highest responsiveness, i.e., BALB/c and C57BL mice, had response curves similar to the C3H/HeJ mice, which had a much lower response. It seemed noteworthy that the early phases of the immune response were significantly different among the strains. On day 3 after immunization, there were essentially no significant differences among the low numbers of PFC in any of the strains tested. However, by day 5 the mouse strains that would eventually show the lowest number of PFC had fewer antibody forming cells than the higher responding strains. Similarly, after the peak response the lower responder strains showed fewer PFC than the other strains.

Since cholera bacilli obviously contain numerous antigenic determinants, it seemed possible that immunocytes may be responding to different antigenic moieties. It has already been suggested that the

Table 2. Cytokinetics of appearance of vibriolytic PFC in spleens of different strains of mice

Time in days after immunization[a]	Mouse strains immunized[b]				
	A/J$^{(a)}$	C57BL/6$^{(b)}$	B6AF$_1^{(a \times b)}$	BALB/c$^{(d)}$	C3H/HeJ$^{(k)}$
0	<5	<5	<5	<5	<5
3	38	37	28	48	20
5	5,180	4,760	1,760	6,500	236
7	11,500	31,400	8,650	49,700	973
10	23,700	72,600	12,400	147,300	1,640
14	32,600	89,100	17,500	175,000	2,750
18	31,400	86,500	11,300	160,000	2,240
22	7,900	64,400	4,200	110,500	1,500
30	2,100	12,300	1,700	62,100	1,100

[a] Spleens from 5–10 mice per group tested for direct IgM vibriolytic PFC on indicated day after immunization.
[b] Average PFC response for indicated mouse group on indicated day after i.p. immunization with 50 μg heat-killed vibrio vaccine.

relationship between the H-2 locus in mice and antibody formation to microbial antigens may be demonstrable only with highly purified bacterial products or antigens with restricted heterogeneity. Therefore, a relationship between immunologic responsiveness to bacteria in the different strains may be more apparent by using suboptimum doses of antigen. Thus groups of inbred mice were injected with graded amounts of cholera bacilli, ranging from 1 to 500 μg per animal. As is evident in Table 3, the general pattern of responsiveness was similar regardless of antigen dose. Although the 50 μg dose resulted in better responses, the high responder mice (i.e., BALB/c) showed essentially similar antibody responses over a 50-fold dose range (i.e., 10 to 500 μg). Similar results were obtained with the other mouse strains, including the low responder C3H animals. When a suboptimum 1.0 μg dose of antigen was used, the "ranking" of the mice to cholera bacilli was still the same, i.e., BALB/c mice showed highest responsiveness and C3H mice the lowest.

Since antibody responses to the type-specific antigen are distinctive for each vibrio serotype and could be readily enumerated in relationship to the antibody response to the common antigen, the cytokinetics of the immune response to these distinctive antigens was also determined (Table 4). In the case of the high responder mice, approximately 10% of the PFC showed specificity to the common antigen during the peak response (day +15 in the experiment shown). Furthermore, on each assay day the number of PFC to the type-specific antigen was much greater than to the common antigen. On the other hand, there was much less difference in the antibody response to the type-specific versus common antigens in the lower responder C3H animals, at least during the peak days of the response. Although in all cases more PFC were evident with specificity to the type-specific antigen, a large percentage of PFC was directed to the common antigen. Similarly, A/J mice, which seemed to be intermediate in their responsiveness to cholera, developed greater numbers of PFC to the common antigen as compared to the type specific antigen. Thus it seems likely that in low or moderate responder animals which do not develop large numbers of PFC to the type-specific antigen there may be a significant response to the common antigen, generally greater than the 1:10 ratio noted in previous studies with high responder animals. This may reflect the ability of mice to respond to and interact with the common antigen in the absence of a maximum immune response to the type-specific antigen. Thus immune responses to these two distinctive antigens on the surface of a single bacterium

Table 3. Effect of dose of cholera vaccine on peak vibriolytic antibody plaque response of different mouse strains

Cholera vaccine dose (μg)[a]	Vibriolytic PFC response[b]				
	A/J$^{(a)}$	C57BL/6$^{(b)}$	B6AF$_1^{(a \times b)}$	BALB/c$^{(d)}$	C3H/HeJ$^{(k)}$
1	2,300	43,000	2,600	97,300	868
10	15,600	72,100	10,350	148,000	1,400
50	29,300	85,500	16,400	160,000	2,450
100	30,100	76,400	10,100	135,000	2,300
500	16,300	42,100	8,200	110,000	2,150

[a] Indicated dose of vaccine injected i.p. into 6–8-week-old mice of indicated strain.
[b] Average splenic PFC response 12–14 days for five to eight mice per group.

Table 4. Cytokinetics of appearance of vibriolytic PFC to distinctive type-specific and common antigens of vibrios in three strains of mice

Time in days after immunization[a]	PFC class[b]	Mouse strain tested[c]		
		A/J[(a)]	BALB/c[(d)]	C3H/HeJ[(k)]
+ 5	A	1,250	1,579	120
	B	4,050	5,700	215
+10	A	7,130	12,650	340
	B	16,400	98,000	1,560
+15	A	9,150	35,000	1,100
	B	20,400	126,000	1,960
+20	A	7,300	18,300	847
	B	12,500	108,400	1,230
+30	A	563	12,200	147
	B	1,700	43,100	833

[a]Groups of mice injected i.p. with 50 μg cholera vaccine on indicated day before testing.

[b]PFC to type-specific antigen B and common antigen A of vibrios detected by vibriolytic plaque assays with homologous and heterologous vibrios in test agar plates.

[c]Average splenic PFC response for five to six mice per group.

may be under separate control, at least in quantitative if not qualitative terms.

One of the advantages of utilizing the cholera model is the absence of pre-existing antibody forming cells before primary immunization. This model therefore permitted an examination of a true primary response in the absence of pre-formed antibody or sensitized immunocytes. In order to compare the results obtained with numerous other studies concerning the genetic control of immune responsiveness to bacterial antigens, comparative studies were performed in which mice were first primed with cholera bacilli before secondary immunization. As is evident in Table 5, many more indirect 7S IgG antibody forming cells appeared in mice given secondary injection of vibrios. There was much less difference in the number of vibriolytic PFC in high responder versus low responder mice after such secondary immunization (Table 5). Late in the primary response BALB/c mice showed many more 7S PFC than did low responder A/J mice. However, after secondary immunization the peak IgG PFC response was about the same in both mouse strains, although early in the response there were more splenic PFC in BALB/c mice. Furthermore, the secondary IgM response was somewhat greater in A/J mice as compared

Table 5. Cytokinetics of appearance of direct (IgM) and indirect (IgG) vibriolytic plaque forming cells in spleens of two mouse strains

Time in days after immunization[a]		Mouse strain tested[b]			
		A/J		BALB/c	
		Direct	Indirect	Direct	Indirect
Primary	+ 5	4,860	0	6,100	0
	+ 10	26,100	258	159,000	565
	+ 15	31,500	2,790	172,000	32,000
	+ 20	12,300	5,760	115,000	63,000
Secondary	0	4,600	2,100	37,600	20,000
	+ 5	65,300	46,000	53,000	75,600
	+ 10	110,200	126,100	87,000	130,500
	+ 15	102,500	114,200	65,000	110,400

[a]Groups of mice immunized by i.p. injection of 50 μg vibrio vaccine either once (primary injection) or twice.

[b]Average PFC response, either direct 19S IgM or indirect 7S IgG, for groups of five to six mice on indicated day after immunization.

to BALB/c mice, indicating that characterization of these mouse strains as high or low responder is not appropriate during the secondary response, which is characterized by the presence of pre-existing PFC and low levels of circulating antibody at the time of booster immunization.

In vitro methods for culturing immunocytes and demonstrating their ability to respond to various antigens in vitro has permitted major advances in understanding control mechanisms of immune responses without the many variable factors complicating the immune system present in intact individuals. Both primary and secondary immunocyte responses to cholera bacilli can be achieved in vitro (26). However, because it is difficult to maintain the viability of mouse immunocytes in vitro, the magnitude of the response of cultured lymphocytes following primary exposure to cholera antigen is much lower than that which is achieved in vivo. This seems to be attributable mainly to the lack of splenocyte viability after 7-10 days in culture, long before the expected PFC response occurs in vivo. Nevertheless, as is evident in Table 6, varying numbers of antibody forming cells developed during the primary immune response of splenocytes from various mouse strains. Furthermore, a strong secondary immune response was readily demonstrable by culturing spleen cells from mice immunized 6-8 weeks earlier with cholera antigens. Differences in the PFC response in vitro, both during the primary and secondary response, seemed to be unrelated to the genetic background of the animals. The primary in vitro response of BALB/c mice was twice as high as that of C3H and A/J mice, which were found in the in vivo studies to be markedly lower. However, spleen cells from C57BL mice, which responded generally much poorer than BALB/c mice in vivo, responded as well if not better than BALB/c spleen cell in vitro. The magnitude of the IgM and IgG PFC responses in vitro by spleen cells from primed animals did not seem to be related to the primary responsiveness of spleen cells in vitro. Splenocytes from BALB/c mice showed the large numbers of direct and indirect PFC in vitro, as did C57BL/6 splenocytes. However, the total number of direct IgM PFC, regardless of mouse strain, was much lower in vitro than in vivo. As occurred in vitro, the range of PFC responses caused by the IgG class of antibody was much narrower than the range for direct IgM PFC. Furthermore, the total number of direct IgM PFC during the secondary response in vitro seemed to be unrelated to the magnitude of the IgG PFC response.

Table 6. Vibriolytic antibody response by spleen cultures from different mouse strains given a primary or secondary immunization in vitro

Mouse strain tested[a]	Vibriolytic PFC response in vitro[b]		
	Primary[c] (direct IgM)	Secondary[c]	
		Direct IgM	Indirect IgG
A/J[(a)]	1,030 ± 276 (6)	8,300 ± 2,100 (7)	12,600 ± 4,500 (4)
C57BL/6J[(b)]	2,960 ± 345 (2)	48,300 ± 2,400 (2)	30,300 ± 2,750 (4)
B6AF$_1$[(a × b)]	1,400 ± 235 (4)	18,500 ± 1,760 (5)	28,100 ± 4,500 (6)
BALB/c[(d)]	2,650 ± 430 (3)	49,300 ± 4,700 (1)	36,300 ± 1,500 (3)
DBA/2[(d)]	976 ± 180 (7)	28,100 ± 3,500 (4)	38,100 ± 2,600 (2)
C3H/HeJ[(k)]	1,300 ± 139 (5)	4,600 ± 953 (6)	16,500 ± 1,380 (7)
C57BL[(k)]	876 ± 38 (8)	15,200 ± 2,100 (6)	30,100 ± 2,600 (5)
NIH	3,852 ± 118 (1)	47,300 ± 4,000 (3)	39,700 ± 3,100 (1)

[a] 5×10^6 spleen cells from indicated mouse strains cultured in vitro.

[b] Average PFC response, either direct (19S IgM) or indirect (7S IgG) 7-8 days after in vitro immunization with 1.0 μg vibrio vaccine; ( ) = ranking of responsiveness.

[c] Spleen cells derived from mice immunized 5-6 weeks earlier by i.p. injection with 50 μg vibrio vaccine.

DISCUSSION AND CONCLUSIONS

It seems apparent from this study that there are marked differences in the magnitude and kinetics of the antibody response to distinctive cell wall antigens of vibrios by spleen cells from nine different strains of mice. No consistent or "standard" number of PFC occurred at any given time interval when mice from different strains were immunized with cholera vaccines. Furthermore, there was no relationship between the magnitude of either the IgM or IgG PFC response during primary versus secondary immune response and the H-2 locus of the mouse strain tested. Strains with a good response to the type-specific antigen often had a disproportionately low response to the common antigen and vice versa. These differences during the primary immune response in vivo, as well as in vitro, could not be related to the effect of persisting sensitized immunocytes or their precursors, or even to pre-existing humoral antibody. There was no natural "background" of immunity, either antibody producing cells or serum antibody, against cholera antigens in any of the mouse strains tested (15–18). It does not seem likely, therefore, that the observed differences in responsiveness among the different mouse strains could reflect varying degrees of pre-sensitization to environmental antigens or be attributable to different "sensitivities" of the individual strains.

During the secondary response, however, it seems likely that the numbers of pre-existing antibody forming cells remaining from the primary immunization could affect not only the magnitude but also the kinetics of the secondary response, both IgM and IgG. Further studies varying the time interval between primary and secondary immunization may be important in examining the mechanism not only of the secondary immune response per se to cholera antigens, but also the genetic control of this response. The availability of a model system in which antibody forming cells can be enumerated at the cellular level to distinctive cell wall antigens of cholera readily permits an examination of the genetic control to specific antigens present on one bacterium. The results obtained suggest that the observed differences in the responsiveness of various inbred strains of mice to these distinctive cholera antigens may reflect a genetic control more complicated than that based on the dosage or activity of specific genes coding for antibody molecules. It is possible that there may be differences in the catabolism or at least recognition of the two cholera antigens, even though they are present in the same bacterial cell. These antigens seem to be distinguished separately, perhaps by specific antigen recognizing cells that function only after the degradation of the bacterium by phagocytic cells. If the common and type-specific antigens were degraded as a unit, most likely similar ratios of antibody forming cells to both antigens should occur in all mouse strains tested.

Experimental results showed a difference in the magnitude of the response to the two different antigens in the various mouse strains examined. Furthermore, the early as well as late stages of antibody formation seemed to be distinguishable during the primary immune response in the different mouse strains. During the secondary response less of a difference was noted for the immune responses to the type-specific versus common antigens. In addition, there was a distinction between the IgG versus IgM antibody responses in the different mouse strains, although these differences were less prominent. Those mouse strains that showed a relatively poor antibody response to the cholera antigens upon primary immunization often showed as good a response as the apparent high responder mice during the secondary response. This could have been attributable to the possibility of different numbers of antigen recognizing or antibody precursor cells present after primary immunization as a function of mouse strain.

It should be noted that attempts were made in this study to determine the susceptibility of various strains of mice to actual infection with viable cholera bacilli. The vibrios grow very rapidly in mice and can be recovered during the first 24 hr or so after injection into animals. However, most strains of mice are quite resistant to mortality by vibrios. At least 10^9 bacteria are required to kill most mice, regardless of strain. However, as indicated earlier, the first antibody producing cells to vibrios do not appear until the second day after immunization, and peak numbers are not present until the end of the second week after immunization. Thus the clearance of vibrios from the circulation and tissue of mice does not seem to be related to antibody forming ability per se. Nevertheless, mice can be readily immunized with appropriate vaccines so as to resist subsequent challenge infection with 10^9 or more organisms. The relationship, however, between antibody formation and actual recovery from sublethal infection remains to be determined. Nevertheless, it seems clear from the studies described here that, at least at the level of antibody forming cells to distinctive vibrio somatic antigens, a genetic control is involved in the response. Additional studies with the same and different strains of mice immunized with other soluble or particulate antigens present on cholera bacilli or extracted from the vibrios may reveal in the future some antibody specificity linked to the H-2 locus. The importance of proper time intervals for determining antibody formation as well as antigen dose are essential in order to determine in more detail the nature of the true primary or secondary response to these antigens in mice.

SUMMARY

The humoral antibody responses to *V. cholera* bacilli is distinctive in that a true primary response characterized by the appearance of large numbers of direct 19S IgM antibody forming cells to both the group and type-specific antigen can be induced. Lower efficiency IgG PFC appear late in the primary response and during the secondary response. Different mouse strains showed distinctive responsiveness, with no correlation between responses to the group versus type-specific antigens. The dose of antigens was important for the magnitude of the immune response, but the kinetics of the responses were generally similar regardless of antigen dose or mouse strain (i.e., peak responses occurred on similar days). The secondary antibody response to vibrios was characterized by appearance of large numbers of 7S IgG PFC; however, the magnitude of this response had little correlation with the primary response. In vitro immune responses, both primary and secondary, were readily induced to the vibrios, with distinctive responses for different mouse strains. There was little relationship between antibody formation in vitro by splenocytes from different mouse strains and responsiveness in vivo. Antibody formation both in vivo and in vitro to cholera bacilli did not correlate with the H-2 locus, suggesting that the genetic control of the immune response to this bacterium was not related to the histocompatibility system.

LITERATURE CITED

1. McDevitt, H. O., and Benacerraf, B. 1969. Genetic control of specific immune responses. Adv. Immunol. 11:31.
2. Benacerraf, B., and McDevitt, H. O. 1972. Histocompatibility linked immune genes. Science 175:273.
3. McDevitt, H. O., and Landy, M. 1972. Genetic Control of Immune Responsiveness. Academic Press, New York.
4. Green, I. 1974. Genetic control of immune responses. Immunogenetics 1:4–21.
5. Benacerraf, B. (ed.). Immunogenetics and Immunodeficiency. University Park Press, Baltimore.
6. Gasser, D. L. 1970. Genetic control of the immune response in mice. J. Immunol. 105:908.
7. Stern, K., Brown, K. S., and Davidson, I. 1956. On the inheritance of natural anti-sheep agglutinin in mice of inbred strains. Genetics 41:517.

8. Ipsen, J. 1959. Differences in primary and secondary immunizability in inbred strains of mice. J. Immunol. 83:448.
9. Gorer, P. A., and Schutze, H. 1938. Genetic studies on immunity in mice. II. Correlation between antibody formation and resistance. J. Hyg. 38:647.
10. Fink, M. A., and Quinn, V. A. 1953. Antibody formation in inbred strains of mice. J. Immunol. 70:61.
11. Robson, H. G., and Vas, S. I. 1972. Resistance of inbred mice to Salmonella typhimurium. J. Inf. Dis. 126:378.
12. Plant, J., and Glynn, A. A. 1976. Genetics of resistance to infection with Salmonella typhimurium in mice. J. Inf. Dis. 133:72.
13. Check, E. W., and Roberts, G. D. 1974. The varying susceptibility of different genetic strains of laboratory mice to histoplasm capsulation. Mycopath. Mycol. Appl. 52:251.
14. Cerny, J., McAlack, R. E., Sajid, M. A., and Friedman, H. 1971. Genetic differences in the immunocyte response of mice to separate determinants on one bacterial antigen. Nature 230:247.
15. McAlack, R. F., Cerny, J., Allan, J. L., and Friedman, H. 1970. Vibriolytic antibody-forming cells: A new application of the Pfeiffer phenomenon. Science 168: 141.
16. Cerny, J., McAlack, R. F., Sajid, M. A., Fronton, J., and Friedman, H. 1971. Early accumulation of antibody plaque-forming cell in mouse spleen lacking a pre-existing immune background. J. Immunol. 106:133.
17. Sajid, M. A., McAlack, R. F., Cerny, J., and Friedman, H. 1971. Antibody plaque responses of mice given Vibrio cholerae antigens as neonates. J. Immunol. 106:1301.
18. Cerny, J., McAlack, R. F., and Friedman, H. 1971. Rapid recruitment and proliferation of antibody plaque-forming cells in the absence of natural antibody. Experientia 27:565.
19. McAlack, R. F., Cerny, J., and Friedman, H. 1971. Cellular formation of vibriolytic antibody by mouse immunocytes: Cytokinetics and specificity of the response. J. Immunol. 107:1752.
20. Cerny, J., McAlack, R. F., and Friedman, H. 1971. Splenic foci of antibody producing cells to two distinct somatic antigens on one bacterium. Immunology 11:445.
21. Cerny, J., McAlack, R. F., and Friedman, H. 1971. Non-random distribution of vibriolytic foci in the spleen of mice lacking "background" antibody. Proc. Soc. Exp. Biol. Med. 137:1021.
22. Cerny, J., McAlack, R. F., and Friedman, H. 1971. An alternative model of antibody forming cell differentiation. Adv. Exp. Med. Biol. 12:323.
23. Kateley, J. R., Patel, C., and Friedman, H. 1974. Detection and enumeration of immunocytes producing different classes of vibriolysins after primary and secondary immunization with cholera vaccine. J. Immunol. 113:1811.
24. Friedman, H. 1975. Vibriolytic IgG immunocyte response of mice after primary and secondary immunization with cholera somatic antigen. Immunology 293:283.
25. Kateley, J. R., and Friedman, H. 1975. The influence of thymus cells on the immune response to cholera somatic and toxin antigen. Ann. N.Y. Acad. Sci. 249:404.

26. Friedman, H., Kamo, I., and Kateley, J. 1975. Cellular events involved in the true primary immune response of splenocytes in vitro. In M. Feldman and A. Globerson (eds.), Immune Reactivity of Lymphocytes, p. 107. Plenum Press, New York.

Infection, Immunity, and Genetics
Edited by Herman Friedman, T. Juhani Linna, and James E. Prier

MULTIGENIC CONTROL OF THE ANTIBODY RESPONSE TO TYPE III PNEUMOCOCCAL POLYSACCHARIDE AND OTHER HELPER T CELL INDEPENDENT ANTIGENS

P. J. Baker, D. F. Amsbaugh, B. Prescott, P. W. Stashak, and J. A. Rudbach

Despite the fact that the antibody response to Type III pneumococcal polysaccharide (SSS-III) is almost monoclonal in nature, the genetic mechanisms that govern immunological responsiveness to this antigen, a linear polymer composed solely of identical disaccharide repeating units (1), are both diverse and complex. They include a gene linked to the X chromosome that determines responsiveness in an almost "all-or-none" manner, as well as autosomal genes that influence the magnitude of the antibody response of mice possessing the X-linked gene (2, 3, 4). An analysis of the effects produced by this second group of "quantitative genes" would shed much light on the mechanisms by which antibody responses and the development of protective immunity are both initiated and controlled; however, this task is made difficult by the fact that at least three interacting cell types, whose activities may be under separate genetic control, participate in the antibody response to SSS-III (5). These cells include: bone marrow-derived precursors of antibody-forming cells (B cells), which synthesize antibody in response to antigen; thymus-derived (T) suppressor cells, which act both directly and indirectly to limit the extent to which B cells proliferate in response to antigen; and amplifier T cells, which act at a later stage during the immune response and, in addition to driving B cells to proliferate further in response to antigen, may also regulate the rate of antibody synthesis by antigen-stimulated B cells (5, 6, 7). The combined activities of all of these cells, therefore, determine the magnitude of the antibody response produced after immunization.

The development of recombinant-inbred (RI) strains of mice, which are produced by inbreeding F_2 progeny from a cross between two genetically distinct progenitor inbred strains (8, 9), enables one to examine these issues in a systematic manner. First, such mice can be used to estimate the number of genes involved in the expression of a given immunological trait. If all RI strains give an immune response identical to that of either progenitor strain in phenotype, a single gene is probably involved. However, the occurrence of unique phenotypes among RI strains indicates the presence of new gene combinations; one can then estimate the *minimal* number of genes involved from the frequency of unique phenotypes observed. Second, some RI strains have been well characterized as to alleles present at a number of different genetic loci (10, 11); this provides a convenient means for determining linkage between genes involved in the expression of any given trait and alleles present at other known loci.

In the present work, RI strains of mice were used to investigate the genetic basis of several factors known to influence the magnitude of the antibody response to SSS-III, dextran B-1355, polyvinylpyrrolidone (PVP), and lipopolysaccharide (LPS) derived from *Escherichia coli* 0113; none of these antigens requires the presence of helper T cells to elicit a normal antibody response in mice (13–16). The results obtained show that neither the ability to give an antibody response to SSS-III nor the expression of suppressor and amplifier T cell activity are linked to genes within the major histocompatibility (H-2) or the IgC_H allotype complex; several autosomal genes, acting in a complementary manner, seem to be involved in the expression of each of these characteristics. With the possible exception of the antibody response to dextran B-1355, which tended to be associated in part with the immunoglobulin heavy chain (IgC_H) allotype complex, genes governing the ability to respond to PVP and LPS were not linked to the H-2 complex, the IgC_H allotype complex, to each other, or to genes associated with responsiveness to SSS-III. Although the antibody response to LPS seemed to be unigenic, several genes were found to influence the magnitude of the antibody response to both dextran B-1355 and PVP.

MATERIALS AND METHODS

Numbers of antibody-producing or plaque-forming cells (PFC), specific for each of the antigens used, provided a measure of the magnitude of the antibody response elicited following immunization with an optimally immunogenic dose of antigen. PFC were detected by a slide version of the technique of localized hemolysis-in-gel (16) using indicator erythrocytes sensitized with SSS-III, palmitoyl-dextran B-1355, LPS, or PVP by established methods (14, 16–18). Only PFC specific for the immunizing antigen are considered, and the results obtained were expressed as the mean number of antigen-specific PFC/spleen for groups of similarly treated mice; Student's *t* test was used to assess the significance of the differences observed. Details concerning the preparation and immunological properties of the antigens used have been reported elsewhere (5, 14, 16–18).

All RI strains of mice, as well as BALB/cBy (C) and C57BL/6By (B) progenitor strains from which they were derived (8, 9) were purchased from The Jackson Laboratory, Bar Harbor, Maine. RI strains were designated as C × BD, C × BE, C × BG, C × BH, C × BI, C × BJ, and C × BK according to existing terminology (8). Information on their H-2 type and IgC_H allotype, relative to that of progenitor C and B mice, has been reported (10, 11). In order to minimize any effects attributed to X-linked genes (2, 3), only female mice (8–12 weeks of age) were used. Male and female progenitor strains gave similar antibody responses in preliminary studies.

RESULTS

Capacity of B Cells to Respond to SSS-III

The magnitude of the antibody response to an optimally immunogenic dose (0.5 μg) of SSS-III was used to evaluate the capacity of B cells to respond to this antigen; this provides a reasonably good index of intrinsic B cell activity, since the magnitude of the antibody response to SSS-III in athymic *nude* and thymus-deprived mice is essentially the same as that of normal intact mice (4, 5).

The data of Table 1 make several points concerning the ability of mice to give a PFC response to an optimally immunogenic dose of SSS-III:

1. Five of seven RI strains examined gave PFC responses that were unique, i.e., the mean values obtained differed significantly in

Table 1. PFC response of RI and progenitor strains of mice to an optimally immunogenic dose (0.5μg) of SSS-III

	PFC response to SSS-III			
Strain	PFC/spleen[a]	Phenotype[b]	H-2 type	IgC$_H$ allotype
B (progenitor)	2.485 ± 0.095 (305) $n = 11$	B	B	B
C × BJ	2.921 ± 0.137 (883) $n = 14$	Unique	B	C
C × BI	3.006 ± 0.191 (1,014) $n = 11$	Unique	B	B
C × BD	3.723 ± 0.075 (5,283) $n = 10$	Unique	C	B
C × BK	3.801 ± 0.073 (6,320) $n = 16$	Unique	B	B
C × BH	3.975 ± 0.091 (9,451) $n = 15$	C	C	B
C (progenitor)	4.102 ± 0.078 (12,654) $n = 11$	C	C	C
C × BE	4.222 ± 0.131 (16,679) $n = 9$	C	B	B
C × BG	4.306 ± 0.039 (20,220) $n = 13$	Unique	B	C

[a] Log_{10} PFC/spleen ± S_x for n mice, 5 days after immunization (i.p.) with 0.5 μg SSS-III; geometric means are in parentheses.
[b] B and C indicate that the response produced was like ($p > 0.05$) that of B or C mice, respectively; unique responses were significantly different ($p < 0.05$) from those of both B and C mice.

magnitude from those of *both* progenitor B and C mice. All of these responses were greater than the response of low-responding B mice. The response of C×BG mice was *greater* than that of high-responding C mice, suggesting overdominance or complementarity between genes. Although the frequency of new or unique phenotypes found suggests that at least five autosomal genes may govern the magnitude of the PFC response to SSS-III, it should be noted that the responses of C × BJ and C × BI mice do not differ ($p > 0.05$) from each other, and the responses of C × BD and C × BK mice do not differ from each other ($p > 0.05$). If one assumes that these pairs share the *same* new combinations of genes, then a minimum of three—rather than five—autosomal genes are involved in determining responsiveness. Obviously, more than one autosomal gene must influence the ability of mice to give a PFC response to SSS-III.

2. These data, as well as those of other studies (2, 20) confirm that the ability of mice to respond to SSS-III is not associated with immune response (*Ir*) genes linked to the H-2 complex. For example, B mice gave the lowest PFC response; yet C × BE and C × BG mice—both of which have the same H-2 type (that of B mice)—produced the highest responses. Also, C × BD and C × BK mice, which differ in H-2 type, gave PFC responses of the same magnitude ($p > 0.05$).
3. The data show that the ability to respond to SSS-III is not linked to genes within the IgC_H allotype complex. For example C × BH and C × BE mice have the same allotype (that of B mice), yet these mice gave responses like those of C mice ($p > 0.05$). In addition, C × BJ and C × BI mice gave similar responses ($p > 0.05$) despite the fact that they differ in allotype. Furthermore, C × BJ and C × BG mice have the same allotype, but gave low responses and high responses, respectively ($p > 0.05$).

The data of Table 2 show once more that B mice and C × BI mice both give low PFC responses to SSS-III. However, mice derived from a cross between these two low responding strains gave a PFC response that was 7–20 times greater than that of either parental strain. Thus, a cross between two low responding strains can give rise to progeny that are high to intermediate responders to SSS-III. Mice derived from a cross between high responding C mice and low responding C × BJ mice produced a PFC response like that of C mice ($p > 0.05$); high responsiveness, rather than low responsiveness, is dominant. These data indicate that the genes that govern the capacity of B cells to respond to SSS-III can act in a complementary manner.

Suppressor T Cell Activity and the Antibody Response to SSS-III

The degree of unresponsiveness produced following pretreatment (priming) with a marginally immunogenic dose (0.005 μg) of SSS-III can be used to determine the amount of suppressor T cell activity present in RI strains of mice; the results of previous studies have shown that this form of unresponsiveness (low-dose paralysis) is an

Table 2. Complementary action of genes governing the ability of mice to give a PFC response to an optimally immunogenic dose (0.5μg) of SSS-III

Strain	PFC/spleen[a]	N mice
B	2.485 ± 0.095 (305)	11
C × BI	3.006 ± 0.191 (1,014)	11
[(C × BI) × B] F_1	3.870 ± 0.081 (7,412)	7
C	4.102 ± 0.078 (12,654)	11
C × BJ	2.921 ± 0.137 (883)	14
[C × (C × BJ)] F_1	4.118 ± 0.093 (13,132)	5

[a]$\text{Log}_{10} \pm S_{\bar{x}}$ for n mice, 5 days after immunization with 0.5 μg SSS-III; geometric means in parentheses.

antigen-specific, T cell dependent phenomenon in which suppressor T cells play an active role (4–6).

The data of Table 3 summarize the results obtained when low-dose paralysis was induced in RI strains of mice; the results are expressed as the *mean percent decrease* in the magnitude of the PFC response to an optimally immunogenic dose (0.5 μg) of SSS-III, following prior treatment with a single injection of 0.005 μg of SSS-III, three days before immunization. B and C mice did not differ significantly ($p > 0.05$) with respect to the degree of low-dose paralysis induced (80–84%), and most strains examined showed the same degree of unresponsiveness, designated by the symbol B/C for phenotype. This was true regardless of H-2 type or IgC_H allotype. Thus, the amount of low-dose paralysis induced, which provides a measure of suppressor T cell activity, is not linked to genes within either of these genetic complexes. More important, the degree of low-dose paralysis induced in C × BI mice was uniquely low, whereas that induced in C × BD mice was uniquely high. The frequency of such new or unique phenotypes, which occurred in two of the seven RI strains examined, implies that at least two autosomal genes may govern the expression of suppressor T cell activity.

Amplifier T Cell Activity and the Antibody Response to SSS-III

The amount of enhancement produced after immunization with an optimally immunogenic dose (0.5 μg) of SSS-III and treatment with antilymphocyte serum (ALS) can be used to assess the influence of amplifier T cells on antigen-stimulated B cells, in the absence of suppressor T cell activity; such enhancement has been found to be T cell dependent and not caused simply by the removal of suppressor T cells or to the stimulation of B cells by ALS (4, 5, 7).

The data of Table 4 show that the amount of enhancement produced following treatment with ALS, which provides a measure of the amount of amplifier T cell activity present, ranged from 6.5- to 190-fold. There seemed to be no direct relationship between the amount of enhancement observed and either IgC_H allotype or H-2 type. With respect to H-2 type, the enhancement obtained with C × BH and C × BD mice was uniquely low; however, the enhancement

Table 3. Degree of low-dose paralysis induced in RI and progenitor strains of mice

Strain	Degree of low-dose paralysis			
	Percent decrease in PFC response[a]	Phenotype[b]	H-2 type	IgC_H allotype
C × BI	49.1 ± 12.3 $n = 9$	Unique	B	B
C × BJ	62.7 ± 11.1 $n = 16$	B/C	B	C
C × BG	70.0 ± 6.5 $n = 12$	B/C	B	C
C × BE	72.7 ± 5.5 $n = 10$	B/C	B	B
B (progenitor)	80.2 ± 7.6 $n = 10$	B/C	B	B
C (progenitor)	84.2 ± 2.9 $n = 11$	B/C	C	C
C × BK	87.3 ± 3.4 $n = 10$	B/C	B	B
C × BH	90.2 ± 2.6 $n = 11$	B/C	C	B
C × BD	94.4 ± 2.0 $n = 13$	Unique	C	B

[a]Mean percent decrease in the PFC response of n mice to 0.5 μg SSS-III, following pretreatment (priming) with 0.005 μg SSS-III. Mice were primed 3 days before immunization with 0.5 μg SSS-III.

[b]B/C indicates that the degree of low-dose paralysis found did not differ ($p > 0.05$) from that induced in either B or C mice. Unique responses were significantly different ($p < 0.05$) from those of both B and C mice.

Table 4. Degree of ALS-induced enhancement of the PFC response to SSS-III in RI and progenitor strains of mice

Strain	Degree of enhancement: Magnitude of increase[a]	Degree of enhancement: Phenotype[b]	H-2 type	IgC_H allotype
C × BH	6.5 ± 1.0 $n = 9$	Unique	C	B
C × BD	8.7 ± 2.3 $n = 10$	Unique	C	B
C × BG	13.4 ± 2.6 $n = 12$	B	B	C
C × BK	14.9 ± 2.2 $n = 11$	B	B	B
C × BE	21.1 ± 0.89 $n = 9$	B	B	B
B (progenitor)	27.1 ± 0.89 $n = 11$	B	B	B
C (progenitor)	82.1 ± 8.1 $n = 11$	C	C	C
C × BJ	119.3 ± 14.4 $n = 10$	Unique	B	C
C × BI	190.7 ± 6.8 $n = 5$	Unique	B	B

[a]Mean increase in the PFC response ± $S_{\bar{x}}$ for n mice given 0.3 ml ALS at the time of immunization with 0.5 μg SSS-III. PFC were assayed 5 days after immunization.

[b]B and C indicate that the enhancement found was like ($p > 0.05$) that of B or C mice, respectively; unique signifies enhancement, significantly different ($p < 0.05$) than that of both B and C mice.

noted for C mice with the same H-2 type was at least 10–15 times greater ($p < 0.05$). The enhancement observed for C × BJ and C × BI mice was uniquely high and in most cases more than five to ten times greater ($p < 0.05$) than that of other strains having the same H-2 type. These unique responses, which occurred in four of the seven RI strains examined, imply that at least three to four autosomal genes may influence the expression of amplifier T cell activity.

Interactions Between Regulatory T Cells Involved in the Antibody Response to SSS-III

In all of the data presented so far, the effects produced by suppressor and amplifier T cells were examined separately. It would be of interest to know if, on the basis of the results shown in Tables 1, 3, and 4, one can make predictions concerning the ability of RI strains to respond to SSS-III. In other words, do mice having low suppressor cell activity and high amplifier cell activity give high antibody responses to SSS-III? The data of Table 5 summarize some of the more pertinent types of cellular interactions observed. Both C × BJ and C × BI mice are low in suppressor activity and high in amplifier activity; one would therefore expect such a combination to result in a high antibody (PFC) response to SSS-III. Instead, both of these strains give low responses. The situation with C × BH and C × BD mice is almost the opposite; one would expect the combination of high suppressor cell activity and low amplifier cell activity to result in low responsiveness, but these strains produce either a high or an intermediate response. These observations suggest that the genes governing the expression of amplifier and suppressor T cell activity interact in a complex—and perhaps competitive—manner, and that such genes may not necessarily operate at the same level of control for the immune response to SSS-III. It should be noted that there is no direct relationship between the ability of B cells to respond to SSS-III and the degree of suppressor activity and amplifier activity expressed, as measured by the correlation coefficient obtained when values for these functions were ranked and compared statistically (Table 6). However, there is an *inverse* relationship between the amount of suppressor and amplifier T cell activity expressed; thus, an *increase* in suppressor activity is associated with a corresponding *decrease* in amplifier activity. This is not surprising in view of the fact that suppressor T cells have been found to inhibit the activities of amplifier T cells as well as B cells (7).

Genetic Control of Antibody Responses to Other Helper T Cell Independent Antigens

RI strains of mice were examined for their ability to give an antibody response to other antigens which—like SSS-III—do not require the presence of helper T cells to give a normal antibody response. The

Table 5. Interrelationships between functional activities[a]

Mice	Suppressor T cell activity	Amplifier T cell activity	Response to SSS-III
C × BJ	Low	High	Low
C × BI	Low	High	Low
C × BH	High	Low	High
C × BD	High	Low	Intermediate

[a]Derived from the data of Tables 1, 3, and 4.

Table 6. Correlation between functional activities[a]

Functional activities compared	Coefficient of correlation[b]	p value
B cell versus suppressor T cell	+ 0.392	> 0.05
B cell versus amplifier T cell	− 0.316	> 0.05
Suppressor T cell versus amplifier T cell	− 0.750	< 0.02

[a]Values listed in Tables 1, 3, and 4 were ranked from the highest to the lowest in order to make the comparisons shown.

[b]Spearman's coefficient of rank correlation (r_s), where r_s is distributed as Student's t with (n-2) degrees of freedom.

data of Table 7 show that the magnitude of the PFC response to an optimally immunogenic dose of LPS derived from *E. coli* 0113 was essentially the same ($p > 0.05$) for all RI and progenitor strains considered. This suggests that the antibody response to this preparation of LPS could be unigenic, which is consistent with the findings of other investigators (20, 21); also, the response produced is not linked to genes within the H-2 or IgC_H allotype complex.

The antibody response to dextran B-1355, which is mainly directed against the α1,3 glucosidic linkage group (12, 13), shows a different pattern (Table 8). Four of the seven RI strains examined gave responses that were unique; this means that more than one autosomal gene must determine responsiveness to this antigen. The unique responses fit into two separate categories; those of C × BK, C × BH, and C × BD mice whose responses do not differ significantly ($p > 0.05$) from each other, and that of C × BG which was in a class by itself. If one assumes that the former group represents mice having the same new combinations of genes, then perhaps three to four autosomal genes may be involved in determining responsiveness to this antigen.

The data of Table 8 also show that the ability to respond to dextran B-1355 is not linked to genes within the H-2 complex. For example C × BK and C × BH mice differ in H-2 type, but give a PFC response of the same magnitude; this was also the case for C × BJ and C mice. C × BH and C mice have the same H-2 type, but differ greatly (> 20-fold) in responsiveness. Although mice having the allotype of B mice tend to give low to intermediate responses to dextran, allotype per se is not the only characteristic associated with responsiveness to this antigen; other factors clearly play an important role and can influence the magnitude of the PFC response of mice having the same allotype by as much as two- to five-fold. For example the responses produced by C × BK, C × BH, and C × BD mice were significantly greater ($p < 0.05$) than that of B mice having the same allotype. In addition, the response of C mice is greater ($p < 0.05$) than that of C × BG mice having the same allotype. To date there is no information concerning the participation of regulatory amplifier and suppressor T cells in the immune response to this antigen.

Table 7. PFC response of RI and progenitor strains of mice to an optimally immunogenic dose (20 μg) of *E. coli* 0113 lipopolysaccharide (LPS)

	PFC response to LPS			
Strain	PFC/spleen[a]	Phenotype[b]	H-2 type	IgC_H allotype
C × BI	3.909 ± 0.038 (8,115) $n = 10$	B/C	B	B
C × BH	3.919 ± 0.035 (8,315) $n = 9$	B/C	C	B
C (progenitor)	3.957 ± 0.039 (9,049) $n = 10$	C	C	C
B (progenitor)	3.969 ± 0.066 (9,330) $n = 9$	B	B	B
C × BG	3.976 ± 0.043 (9,467) $n = 10$	B/C	B	C
C × BD	4.061 ± 0.058 (11,498) $n = 4$	B/C	C	B
C × BE	4.039 ± 0.056 (11,927) $n = 10$	B/C	B	B
C × BK	4.087 ± 0.071 (12,224) $n = 8$	B/C	B	B
C × BJ	4.155 ± 0.060 (14, 274) $n = 6$	B/C	B	C

[a] Log_{10} PFC/spleen ± S_x for n mice, 4 days after immunization (i.v.) with 20 μg LPS; geometric means are in parentheses.

[b] B/C signifies that the response obtained did not differ significantly ($p > 0.05$) from that of either B or C mice.

Table 8. PFC response of RI and progenitor strains of mice to an optimally immunogenic dose (100 μg) of dextran B-1355

	PFC response to dextran			
Strain	PFC/spleen[a]	Phenotype[b]	H-2 type	IgC_H allotype
B (progenitor)	2.478 ± 0.132 (301) $n = 10$	B	B	B
C × BI	2.890 ± 0.261 (776) $n = 15$	B	B	B
C × BE	2.909 ± 0.250 (812) $n = 15$	B	B	B
C × BK	3.073 ± 0.161 (1,183) $n = 8$	Unique	B	B
C × BH	3.116 ± 0.174 (1,305) $n = 13$	Unique	C	B
C × BD	3.167 ± 0.143 (1,469) $n = 13$	Unique	C	B
C × BG	4.193 ± 0.089 (15,609) $n = 9$	Unique	B	C
C × BJ	4.324 ± 0.106 (21,083) $n = 10$	C	B	C
C (progenitor)	4.448 ± 0.053 (28,030) $n = 19$	C	C	C

[a] Log_{10} PFC/spleen ± $S_{\bar{x}}$ for n mice, 5 days after immunization with 100 μg dextran B-1355; geometric means are in parentheses.

[b] B and C signify that the response produced was like ($p > 0.05$) that of B or C mice, respectively; unique responses differed significantly ($p < 0.05$) from those of both B and C mice.

The results obtained with PVP are summarized in Table 9. There is no direct relationship between the ability to respond to PVP and either H-2 type or IgC_H allotype. The response of C mice was at least five times greater than that of C × BD mice having the same H-2 type, whereas the response of B mice was less than that of both C × BK and C × BE mice having the same H-2 type. C × BD and C × BE mice share the same IgC_H allotype, that of B mice, yet these mice

Table 9. PFC response of RI and progenitor strains of mice to an optimally immunogenic dose (0.25 μg) of PVP-K90

Strain	PFC response to PVP: PFC/spleen[a]	Phenotype[b]	H-2 type	IgC_H allotype
C × BD	3.408 ± 0.033 (2,532) $n = 8$	B	C	B
B (progenitor)	3.502 ± 0.085 (3,176) $n = 10$	B	B	B
C × BI	3.513 ± 0.086 (3,257) $n = 4$	B	B	B
C × BJ	3.786 ± 0.061 (6,109) $n = 13$	Unique	B	C
C × BG	3.806 ± 0.055 (6,392) $n = 11$	Unique	B	C
C × BH	3.899 ± 0.097 (7,933) $n = 10$	C	C	B
C × BK	4.004 ± 0.064 (10,104) $n = 12$	C	B	B
C (progenitor)	4.108 ± 0.070 (12,827) $n = 10$	C	C	C
C × BE	4.113 ± 0.117 (12,980) $n = 6$	C	B	B

[a]Lof_{10} PFC/spleen ± $S_{\bar{x}}$ for n mice, 5 days after immunization with 0.25 μg PVP; geometric means are in parentheses.

[b]B and C indicates that the response produced was like ($p > 0.05$) that of B or C mice, respectively; unique responses were significantly different ($p < 0.05$) from those of both B and C mice.

produced the lowest and highest responses found, respectively. The responses produced by C × BJ and C × BG mice were unique and of the same magnitude; this suggests that these phenotypes may represent mice possessing the same new combinations of genes determining responsiveness to PVP and that perhaps as many as two to three autosomal genes may govern responsiveness to this antigen. Although amplifier and suppressor T cells have been shown to participate in the antibody response to PVP (14, 22), RI strains of mice have not as yet been examined for these activities as they related to the response to PVP.

Strain Distribution Patterns of Phenotypes

Phenotypes for RI strains immunized with dextran B-1355, PVP, *E. coli* LPS or SSS-III are summarized in Table 10. From the strain distribution patterns shown, one may make the following statements concerning the ability of mice to make an antibody response to these antigens:

1. With the possible exception of the antibody response to dextran, which is associated *in part* with IgC_H allotype, genes governing the ability to respond to each of the remaining antigens do not seem to be linked to H-2 type, to IgC_H allotype, or to each other. To date there is no evidence to indicate that any of the *Ir* genes involved are on the same chromosome or are linked to any known genetic trait. Studies with other antigens and different RI strains would provide useful information in this regard.
2. Although several unique responses have been noted, they are not restricted to any particular RI strain, nor is a unique response to one antigen associated with a unique response to any other antigen. Because all RI strains gave similar ($p > 0.05$) responses to LPS, these findings suggest that the differences found are most likely a reflection of genetic differences in the ability to respond to a particular antigen, rather than more general nonspecific factors such as spleen size, body weight, absolute number, and/or functional capacity of B cells, etc.
3. There is no direct relationship between the number of unique responses found, which is characteristic of both the complexity and number of gene interactions involved in determining responsiveness, and the structure and/or chemical composition of antigen per se. The structure and chemical composition of *E. coli* LPS (no unique responses) is certainly much more diverse and complex than that of dextran (a branched glucose polymer), SSS-III (a linear polymer of glucose—glucuronic acid), and PVP (a linear polymer of vinylpyrrolidone monomers).

The present work illustrates that *Ir* genes not linked to the H-2 or IgC_H allotype complex play a significant role in determining immunological responsiveness to several antigens that elicit highly restricted antibody responses; such genetic control, which often involves the participation of several genes, is expressed at both the B

Table 10. Strain distribution patterns of phenotypes for antibody responses to four different antigens in RI strains of mice

RI strain	Antigen			
	SSS-III	Dextran B-1355	PVP	*E. coli* 0113 LPS
C × BD	Unique[a]	Unique	B	B/C
C × BE	C	B	C	B/C
C × BG	Unique	Unique	Unique	B/C
C × BH	C	Unique	C	B/C
C × BI	Unique	B	B	B/C
C × BJ	Unique	C	Unique	B/C
C × BK	Unique	Unique	C	B/C

[a]B or C indicate that the response obtained was like ($p > 0.05$) that of progenitor B or C mice, respectively. Unique responses were significantly different ($p < 0.05$) from those of both B and C mice. B/C indicates that the responses obtained did not differ significantly ($p > 0.05$) from those of either B or C mice.

cell and the regulatory T cell (amplifier and suppressor cell) level. The extent to which these genetic mechanisms influence immune responses to other antigens that are under H-2-linked genetic control and their relationship to the development of protective immunity remain to be evaluated.

LITERATURE CITED

1. Howe, M. J., Brimacombe, J. S., and Stacey, M. 1964. The pneumococcal polysaccharides. Adv. Carbohy. Res. 19:303.
2. Amsbaugh, D. F., Hansen, C. T., Prescott, B., Stashak, P. W., Barthold, D. R., and Baker, P. J. 1972. Genetic control of the antibody response to Type III pneumococcal polysaccharide in mice. I. Evidence that an X-linked gene plays a decisive role in determining responsiveness. J. Exp. Med. 136: 931.
3. Amsbaugh, D. F., Hansen, C. T., Prescott, B., Stashak, P. W., Asofsky, R., and Baker, P. J. 1974. Genetic control of the antibody response to Type III pneumococcal polysaccharide in mice. II. Relationship between IgM immunoglobulin levels and the ability to give an IgM antibody response. J. Exp. Med. 139:1499.
4. Baker, P. J., Amsbaugh, D. F., Prescott, B., and Stashak, P. W. 1976. Genetic control of the antibody response to Type III pneumococcal polysaccharide in mice. III. Analysis of genes governing the expression of regulatory T cell activity. J. Immunogenet. 3:275.
5. Baker, P. J. 1975. Homeostatic control of antibody responses: A model based on the recognition of cell-associated antibody by regulatory T cells. Transplant. Rev. 26:3.

6. Markham, R. B., Stashak, P. W., Prescott, B., Amsbaugh, D. F., and Baker, P. J. 1977. Effect of Concanavalin A on lymphocyte interactions involved in the antibody response to Type III pneumococcal polysaccharide. I. Comparison of the suppression induced by Con A and low-dose paralysis. J. Immunol. 118:952.
7. Markham, R. B., Reed, N. D., Stashak, P. W., Prescott, B., Amsbaugh, D. F., and Baker, P. J. 1977. Effect of Concanavalin A on lymphocyte interactions involved in the antibody response to Type III pneumococcal polysaccharide. II. Ability of suppressor T cells to act on both B cells and amplifier T cells to limit the magnitude of the antibody response. J. Immunol. 119:1163.
8. Bailey, D. W. 1971. Recombinant-inbred strains. Transplantation 11:325.
9. Swank, R. T., and Bailey, D. W. 1973. Recombinant-inbred lines: Value in genetic analysis of biochemical variants. Science 181:1249.
10. Potter, M., Finlayson, J. S., Bailey, D. W., Muschinski, E. B., Reamer, B. L., and Walters, J. L. 1973. Major urinary protein and immunoglobulin allotypes of recombinant-inbred mouse strains. Genet. Res. 22:325.
11. Potter, M., Pumphrey, J. G., and Bailey, D. W. 1975. Genetics of susceptibility to plasmacytoma induction. I. BALB/c AnN (C), C57BL/6N (B6), C57BL/Ka (BK), (C × B6)F_1, (C × BK)F_1 and C × B recombinant-inbred strains. J. Nat. Cancer Inst. 54:1413.
12. Baker, P. J., Reed, N. D., Stashak, P. W., Amsbaugh, D. F., and Prescott, B. 1973. Regulation of the antibody response to Type III pneumococcal polysaccharide. I. Nature of regulatory cells. J. Exp. Med. 137:1431.
13. Howard, J. G., and Courtenay, B. M. 1975. Influence of molecular structure on the tolerogenicity of bacterial dextrans. II. The α1,3 linked epitope of dextran B-1355. Immunology 29:599.
14. Lake, J. P., and Reed, N. D. 1976. Regulation of the immune response to polyvinyl pyrrolidone: Effect of antilymphocyte serum on the response of normal and nude mice. Cell. Immunol. 21:364.
15. Manning, J. K., Reed, N. D., and Jutila, J. W. 1972. Antibody response to *Escherichia coli* lipopolysaccharide and Type III pneumococcal polysaccharide by congenitally thymusless (nude) mice. J. Immunol. 108:1470.
16. Baker, P. J., Stashak, P. W., and Prescott, B. 1969. Use of erythrocytes sensitized with purified pneumococcal polysaccharide for the assay of antibody and antibody-producing cells. Appl. Microbiol. 17:422.
17. Leon, M. A., Young, N. M., and McIntire, K. R. 1970. Immunochemical studies of the reaction between a mouse myeloma macroglobulin and dextrans. Biochemistry 9:1023.
18. Rudbach, J. A. 1971. Molecular immunogenicity of bacterial lipopolysaccharide antigens: Establishing a quantitative system. J. Immunol. 109:993.
19. Braley, H. C., and Freeman, M. J. 1971. Strain differences in the antibody plaque-forming cell responses of inbred mice to pneumococcal polysaccharide. Cell. Immunol. 2:73.
20. DiPauli, R. 1973. Genetics of the immune response. II. Inheritance

of antibody specificity to lipopolysaccharides in mice. J. Immunol. 111:82.

21. Watson, J., and Riblet, R. 1974. Genetic control of responses to bacterial lipopolysaccharides in mice. I. Evidence for a single gene that influences mitogenic and immunogenic responses to lipopolysaccharides. J. Exp. Med. 140:1147.
22. Lake, J. R., and Reed, N. D. 1976. Characterization of antigen-specific immunologic paralysis induced by a single low dose of polyvinylpyrrolidone. J. Retic. Soc. 20:307.

Infection, Immunity, and Genetics
Edited by Herman Friedman, T. Juhani Linna, and James E. Prier

PROBLEMS IN THE EXPERIMENTAL ANALYSIS OF CELL-MEDIATED IMMUNITY IN VIRUS INFECTIONS

P. C. Doherty

Perhaps one of the least well understood aspects of virus pathogenesis concerns the nature and specificity of the cell-mediated immune (CMI) response (1, 2). The reasons for this are at least threefold:

1. Suitable in vitro assays for analyzing CMI have only become available over the past three or four years, being still only well characterized for mouse model systems.
2. Use of these in vitro techniques has led to the realization that immune thymus-derived lymphocytes (T cells), the central component of CMI (3), show specificity for both the virus used to immunize and for the major histocompatibility genes (*H-2* genes in the mouse) expressed in the sensitizing environment (1, 2). This finding has raised fundamental questions concerning the nature of both T cell recognition in general, and of the relevant antigenic changes on virus-infected cells in particular.
3. The discovery that *H-2* genes are involved in T cell responses in syngeneic systems has led to the concept that the triggering event in CMI is essentially a cell-interaction process. This involves, on the one hand, specific receptor(s) expressed on the plasma membrane of the immune lymphocyte and, on the other, surface characteristics of target cells determined by both *H-2* genes and by the viral genome. Rigorous analysis of such a transient interaction between two diverse cell types is not readily achieved.

We may add to these specific problems stated for the virus models several further difficulties in the overall definition of T cell recognition. These persist despite many years of intensive investigation of allogeneic and helper (especially using hapten-carrier protocols) T cell systems (4, 5). At present there are no continuously growing clones of functionally specific T cells available. Analysis of the binding site on the T cell receptor thus presents a considerable problem, although significant progress may have been made recently (6, 7). Furthermore, the basic structure of the T cell recognition unit is still a matter of debate. We do not even know whether (relative to an antibody molecule) it is large or small. In fact it is not at present clear whether individual T cells express one, two, or even more than two different receptors (8–11).

THE ANTIGENS RECOGNIZED BY VIRUS-IMMUNE T CELLS

Current debate (1, 2, 8–10) is concerned basically with whether "Virus-immune" T cells express two different variable region gene (*V* gene) products, the one specific for self *H-2* and the other for viral antigen, or a single receptor binding to an "altered self" component (12). "Altered self" models propose variously that the recognition unit may interact with a "junction zone" between viral and *H-2* antigens, with specific carbohydrate patterns on the virus that are in some way dictated by *H-2* genes, or with some form of "derepressed" alloantigen (1, 13, 14). The diversity of these speculations indicates the present state of knowledge.

The alternative "two *V* gene" model incorporates a major heresy "self markers" are involved in immune recognition. However, it i otherwise more conservative insofar as recognition of viral antigen per se is considered to occur. Even if this is shown to be true, we still need to know much more about which viral components are involved Resolution of such questions may have considerable implications fo the further development of rational vaccination procedures.

The scope of the problem may be stated in the context of current experiments with type A influenza viruses (15–17). It seems that a major component of the cytotoxic T cell response generated in mice immunized with influenza A viruses is highly cross-reactive for cells infected with all type A, but not with type B, influenza viruses. The recognition spectrum of this particular T cell subset thus differs totally from that defined by serological analysis of the surface hemagglutinin (H) or neuraminidase (N) antigens expressed on the virion (18). This observation raises several possibilities. The T cell receptor may be specific for some form of "altered self" antigen, common expression of which is induced by viruses of similar molecular biology. Alternatively, recognition may be directed at shared components on the H o

N molecules, which are not normally "seen" serologically during infection. A further idea, for which some evidence now exists (19, 20), is that the internal matrix (M) and ribonucleoprotein (RNP) antigens shared by all type A influenza viruses may be expressed on cell surface.

Resolution of this problem may not be simple. Blocking of effector T cell function with anti-viral antibodies is not readily achieved (21). Attempts at stimulating cytotoxic T cells with various subviral components have given equivocal results (22). Furthermore, specific binding of free antigen to immune T cells has not been satisfactorily demonstrated.

Some approaches are obvious. It seems apparent that we need to look much more closely at the surface characteristics of virus-infected cells, rather than simply concentrating on the properties of free virus particles. Emphasis is given to this point by the fact that cytotoxic T cells can be shown to interact with cells replicating Coxsackie viruses (23), even though infection with such viruses is not known to induce specific modification of the cell plasma membrane. The alternative is to attempt rigorous analysis of the T cell receptor with anti-idiotype antibodies (6, 7). Rational experiments along these lines presuppose, however, that we have some idea of which viral antigens are recognized by "virus-immune" T cells.

PROTECTION, IMMUNOPATHOLOGY, AND AUTOIMMUNITY

Experiments with mice have established that in some situations T cell responses may be protective, in others deleterious. The classic example of protection is the ectromelia (mouse pox) model (3, 24), in which a lytic virus growing in liver and spleen is eliminated by sensitized T cells and macrophages acting in concert. Immunopathological process is exemplified by the lymphocytic choriomeningitis system (25-28), where T cell-mediated lysis of ependymal and choroid plexus cells infected with a non-lytic virus may result in death of the host. It seems probable that both protective and pathological aspects of CMI co-exist in most virus infections. We need to know, however, if rational therapeutic (e.g., immunosuppression) and vaccination procedures are to be developed, which situation is likely to be of greatest importance in a particular disease process.

Our ignorance in this area is considerable. T cells are known to be effective (together with macrophages) in eliminating virus from some solid tissues. However, we have no clear idea of how well T cells function at body surfaces, such as the respiratory and gut epithelium.

Little information is available concerning T cell invasion, or effectiveness, in brain parenchyma. The nature of the stimulus leading to T cell extravasation from blood is not understood, and there is a paucity of knowledge about the functional components (other than T cells and activated macrophages) of virus-induced inflammatory exudates. Are "natural killer" cells operating (29–31)? Are antibody-dependent cell-mediated immune mechanisms (32–33) functional in the site of pathology?

Several virus infections are known to be associated with early, non-specific proliferation of T cell clones which are reactive to normal target cells (34, 35). What are the possible long-term consequences of introducing such bystander cells into, say, brain during the process of acute T cell-mediated virus clearance?

The idea that some form of "altered self" antigen may be recognized by virus-immune T cells raises the possibility that infection with quite common viruses (e.g., measles) may lead to sensitization against cross-reactive host components normally sequestered in cryptic sites (e.g., brain). Furthermore, it is now apparent that sequential infection with serologically distinct influenza A viruses leads to second set responses of much greater magnitude than that seen in primary infections (16). Recurrent bursts of T cell stimulation may thus result (at least in mice) from repeated exposure to similar viruses. The same situation may apply with influenza B viruses, rhinoviruses, and adenoviruses (although we have no information on this). Cyclical occurrence of immunopathological process could thus occur if there is either cross-reactivity with self, or if viral antigen persists in some form in sites such as brain.

T CELL REGULATION

Mechanisms by which immune T cell responses are regulated, and homeostasis is thus maintained, are a current obsession of cellular immunologists. This situation is likely to continue, because the various suppressor and helper (T and B or T and T) phenomena seem to be extremely complex (36, 37). Such information may be of eventual relevance to manipulation of T cell responses in virus systems, although little specific knowledge exists at present with regard to infectious disease models. A similar statement can be made concerning immune response (*Ir*) gene effects (38).

Insofar as cytotoxic (surveillance) T cell function is concerned, however, the various virus assays probably offer the best available technology for fundamental immunological analysis (1). Infection with a virus is a much more biological event than, say, coating a cell with TNP (39). Furthermore, the in vivo virus models are both readily manipulated and unique in immunology (1, 3). Studies of infectious process may thus continue to contribute generally to our overall understanding of cellular interactions in the regulation of immune responsiveness.

CONCLUSIONS

The nature of cell-mediated immunity in virus infections is beginning to be defined in cellular terms for one species, the mouse. Very little rigorously stated information exists for other animal models, or for man (1, 2). Analysis at a molecular level is likely to be difficult, because we are considering transient interactions between diverse cell types. Experimental approaches to the dual problems of cell surface-antigenicity and the T cell recognition structure(s) would seem, however, to be feasible. The various virus systems offer unique models for the fundamental analysis of immunological and pathological process. Such information is essential for the further development of rational vaccination and therapeutic procedures, and it should contribute to our general biological understanding.

LITERATURE CITED

1. Doherty, P. C., Blanden, R. V., and Zinkernagel, R. M. 1976. Transplant. Rev. 29:84.
2. Zinkernagel, R. M., and Doherty, P. C. 1977. Contemporary Topics in Immunobiology 7:179.
3. Blanden, R. V. 1976. Transplant. Rev. 19, 56.
4. Cerottini, J. -C. and Brunner, K. T. 1974. Advances in Immunology 19. 67.
5. Janeway, C. A., Jr. 1976. Transplant. Rev. 29, 164.
6. Doherty, P. C., Götze, D., Trinchieri, G. and Zinkernagel, R. M. 1976. Immunogenetics 3, 517.
7. Zinkernagel, R. M. and Doherty, P. C. 1976. Cold Spring Harbor Symp. Quant. Biol. 41, 505.
8. Binz, H. and Wigzell, H. 1975. J. Exp. Med. 142, 1218.
9. Black, S. J., Hämmerling, G., Berek, C., Rajewsky, K. and Eichmann, K. 1976. J. Exp. Med. 143:846.
10. Janeway, C. A., Jr., Binz, H. and Wigzell, H. 1976. Scand. J. Immunol. 5, 993.
11. Lindahl, K. F. and Wilson, D. B. 1977. J. Exp. Med. 145, 508.
12. Zinkernagel, R. M. and Doherty, P. C. 1974. Nature (Lond.) 251, 547.
13. Blanden, R. V., Hapel, A. J. and Jackson, D. C. 1976. Immunochemistry 13, 179.
14. Garrido, F., Schirrmacher, V. and Festenstein, H. 1976. Nature (Lond.) 259. 228.

15. Effros, R. B., Doherty, P. C., Gerhard, W. and Bennink, J. 1977. J. Exp. Med. 145, 557.
16. Doherty, P. C., Effros, R. B. and Bennink, J. 1977. Proc. Natl. Acad. Sci. USA 74, 1209.
17. Zweerink, H., Courtneidge, S. A., Skehel, J. J., Crumpton, M. J. and Askonas, B. A. 1977. Nature (Lond.) 267, 354.
18. Schild, G. C. and Dowdle, W. R. 1975. In "The Influenza Viruses and Influenza," E. D. Kilbourne, ed., p. 315.
19. Biddison, W. E., Doherty, P. C. and Webster, R. G. 1977. J. Exp. Med. 146, 690.
20. Virlizier, J. L., Allison, A. C., Oxford, J. S. and Schild, G. C. 1977. Nature (Lond.) 266, 52.
21. Blanden, R. V., Pang, T. E. and Dunlop, M. B. C. 1977. In "Cell Surface Reviews," Vol. 2, G. Poste and G. L. Nicolson, eds., p. 249.
22. Shellam, G. R., Knight, R. A., Mitchison, N. A., Gorczynski, R. M. and Maoz, A. 1976. Transplant. Rev. 29, 249.
23. Wong, C. V., Woodruff, J. J. and Woodruff, J. F. 1977. J. Immunol. 118, 1159.
24. Kees, U. and Blanden, R. V. 1976. J. Exp. Med. 143, 450.
25. Volkert, M., Marker, O. and Bro-Jorgensen, K. 1974. J. Exp. Med. 138, 1266.
26. Cole, G. A., Nathanson, N. and Prendergast, R. A. 1972. Nature (Lond.) 238:335.
27. Doherty, P. C. and Zinkernagel, R. M. 1974. Transplant. Rev. 19, 89.
28. Doherty, P. C., Dunlop, M. B. C., Parish, C. R. and Zinkernagel, R. M. 1976. J. Immunol. 117, 187.
29. Kiessling, R., Klein, E. and Wigzell, H. 1975. Eur, J. Immunol. 5, 112.
30. Herberman, R. B., Nunn, M. E., Holden, H. T. and Lavrin, D. H. 1975. Int. J. Cancer 16, 230.
31. Wolfe, S. A., Tracey, D. E. and Henney, C. S. 1976. Nature (Lond.) 262, 584.
32. Perlmann, P., Perlmann, H. and Wigzell, H. 1972. Transplant. Rev. 13, 91.
33. Maclennan, I. C. M. 1972. Transplant. Rev. 13, 67.
34. Gardner, I., Bowern, N. A. and Blanden, R. V. 1974. Eur. J. Immunol. 4, 63.
35. Pfizenmaier, K., Trostmann, H., Röllinghoff, M. and Wagner, H. 1975. Nature (Lond.) 258, 238.
36. Gershon, R. K. 1975. Transplant. Rev. 26, 170.
37. Katz, D. H. and Benacerraf, B. 1975. Transplant. Rev. 22, 175.
38. McDevitt, H. O., Oldstone, M. B. A. and Pincuss, T. 1974. Transplant. Rev. 19, 209.
39. Shearer, G. M. 1974. Eur. J. Immunol. 4, 257.

Infection, Immunity, and Genetics
Edited by Herman Friedman, T. Juhani Linna, and James E. Prier

GENETIC CONTROL OF THE EFFECTS OF LEUKEMIA VIRUSES ON IMMUNE RESPONSES

Steven C. Specter, Walter S. Ceglowski, and Herman Friedman

The susceptibility of mice to infection by a murine leukemia virus such as Friend leukemia virus (FLV) seems to be controlled by several genes (1–4). The major histocompatability complex (MHC), or the *H-2* locus of the mouse has been found to control the susceptibility of mice to leukemia viruses (3–4). Resistance of mice to FLV infection seems to be associated with the genotype $H\text{-}2^b$, whereas mice of genotype $H\text{-}2^d$ are susceptible. The genetic dominance of susceptibility is evident since the heterozygote $H\text{-}2^{b/d}$ is always susceptible to leukemogenesis by such viruses (3). However, it should be noted that even more restrictions within the MHC than reported in previous studies have been observed. For example, Chesebro, Wehrly, and Stimfling (3) reported that the D region of this locus is responsible for susceptibility and they designated a single gene, *RFV-1*, for recovery from Friend virus disease (3). However, as Lilly recently indicated (1), the D region is not universally recognized as the most important segment of the *H-2* locus regulating susceptibility to FLV. Since congenic F_1 $H\text{-}2^b$ mice are more likely to recover than parental $H\text{-}2^b$ mice from FLV infection, other genes besides *RFV-1* seem to be involved in resistance and recovery from virus leukemogenesis in mice (3).

Two other genes besides the *H-2* gene, designated *Fv-1* and *Fv-2*, have been associated with susceptibility to murine leukemogenesis (1,2). FLV is now considered to be a complex of at least two viruses, i.e., the lymphatic leukemia virus (LLV), which is considered a helper virus, and the spleen focus forming virus (SFFV), considered defective and requiring LLV for replication (5). Susceptibility to LLV is regulated by the *Fv-1* gene. Alleles are designated $Fv\text{-}1^n$ and $Fv\text{-}1^b$. These terms refer to the tendency of the virus to be N-tropic (i.e., infect NIH mice), or B-tropic (i.e., infect BALB/c mice). Some viruses are

N-B-tropic and can infect mice of either type (1). Susceptibility to SFFV is thought to be controlled by the *Fv-2* gene, with alleles referred to as *Fv-2^s^* or *FV-2^r^* for susceptibility (the dominant gene) or resistance, respectively (1).

Although there are many studies concerning genetic control of susceptibility to murine virus-induced leukemogenesis, including that induced by FLV, much less attention has been given to the role of genetic factors controlling leukemia virus-induced immunosuppression. It is now widely recognized that FLV, as well as many other oncornaviruses and some non-oncogenic viruses, may induce varying degrees of immunosuppression in susceptible as well as resistant strains of mice (6). A control mechanism involving the *Ir* gene in the MHC and genetically regulated susceptibility to FLV-associated immune suppression has been postulated (1,7). However, there is little evidence as yet to support this hypothesis. Furthermore, susceptibility to immunological impairment by FLV may not be under direct genetic control but rather may be a consequence of genetic susceptibility to the leukemogenic process per se which, in turn, results in a depressed immune competence by some other mechanism. In this regard, studies in this laboratory over the past decade have been concerned with the mechanism of suppression induced by murine leukemia viruses such as FLV, both at the cellular and humoral levels. Results of such studies indicate that mice genetically susceptible or resistant to FLV may evince varying degrees of impairment of cellular and/or humoral immune responses as a consequence of exposure to this tumor virus.

GENERAL EXPERIMENTAL DESIGN AND PROCEDURES

BALB/c ($H\text{-}2^d$, $Fv\text{-}1^b$, $Fv\text{-}2^s$) and C57BL/6 ($H\text{-}2^b$, $Fv\text{-}1^b$, $Fv\text{-}2^r$) mice were injected by intraperitoneal (i.p.) inoculation with graded doses of FLV and examined at various times thereafter for either humoral or cellular immune responses (2,7). In terms of humoral immunity, mice were immunized with sheep red blood cells (SRBC) and tested for splenic antibody plaque forming cells (PFC) to the erthrocytes by both the direct and indirect hemolytic plaque assays for primary or secondary antibody responses (8,9). Serum was collected from mice and tested for hemagglutinating and hemolytic antibodies. For secondary challenge injection mice were given SRBC 3 weeks after initial immunization with SRBC and examined specifically for both direct IgM and indirect 7S IgG PFC by facilitation procedures with anti-IgG serum. Assessment of cellular immunocompetence was performed using macrophage migration inhibition (MI) assays with splenocytes and peritoneal exudate cells from infected and control mice (10,11). In addition, lymphocyte-mediated cytolysis (LMC) of allogeneic target tumor cells was also used as a measure of cellular immunity in leukemia virus infected mice (12,13). In all cases immune responses were determined after exposure of susceptible and resistant mouse strains to the virus as compared to matched control mice treated in the same manner, except that they were either not exposed to virus or were given inactivated virus preparations.

EXPERIMENTAL RESULTS

Assessment of Humoral Immunity in Susceptible and Resistant Mice

Immune suppression of the primary humoral antibody response to SRBC was noted in assays for both the hemagglutination (HA) titer and the PFC response in BALB/c mice within one day after infection. The impairment of immunity was progressive and 7 days after infection PFC numbers decreased more than 99% and serum titers declined at least fourfold as compared to responses of control non-infected animals (Table 1). However, it should be noted that for leukemia resistant C57BL mice injected with the same preparation of FLV, immunosuppresion was also noted 1 and 3 days after virus administration. By the seventh day after virus injection, both PFC numbers and HA titers returned to normal. Immunosuppression was directly correlated with the recovery of virus from spleens of the C57BL mice. When spleen suspensions were positive for the presence of FLV, immunosuppression was evident. On the other hand, when virus was no longer detectable in the spleen of virus-infected mice, immune competence was fully restored (Table 1).

A marked suppression of the secondary humoral immune response to SRBC was also readily demonstrable in susceptible BALB/c mice after infection with FLV. Both direct IgM and indirect IgG PFC responses were inhibited. As shown in Table 2, leukemia resistant C57BL mice were found to be transiently immunosuppressed with respect to direct IgM PFC during the secondary response, with full restoration of immune reactivity 7 days post-infection. However, the numbers of indirect PFC were still greatly reduced even at this time after FLV infection; a time when virus was no longer detectable in the spleen. Nevertheless, reversal of immunosuppression seemed to be occurring even at this time, i.e., there was a higher response 7 days

Table 1. Effect of leukemia virus infection on the hemolytic antibody response of BALB/c and C57BL mice to sheep red cells

Time between infection and SRBC challenge[a] (days)	BALB/c			C57BL/6			
	PFC/spleen[b]	Percent of control	Mean serum titer	PFC/spleen	Percent of control	Mean serum titer	Virus presence
Controls (no FLV)	75,100		1:128	56,700		1:96	Negative
−1	12,000	16	1:48	6,800	12	1:46	Positive
−3	3,750	5	1:32	2,288	4	1:24	Positive
−7	585	1	1:16	62,000	106	1:96	Negative
−14				58,400	103		Negative

[a]Mice injected i.p. with FLV on indicated day before immunization with SRBC; splenic PFC tested four days later.

[b]Spleens were homogenized, clarified, and 0.5 ml of a 20% homogenate suspension was inoculated into susceptible BALB/c mice; presence of splenomegaly 28 days after inoculation was considered positive for virus recovery.

Table 2. Secondary immune responses to SRBC in FLV infected C57BL/6 mice

Day of FLV inoculation relative to day of secondary immunization[a]	Antibody response/spleen[b]			
	Direct (IgM)	Percent of control	Indirect (IgG)	Percent of control
Controls (uninjected)	2.7×10^4		1.9×10^5	
+1	2.8×10^4	103	1.6×10^5	84
0	3.0×10^4	110	1.8×10^5	94
−1	1.7×10^4	63	8.4×10^4	44
−3	2.0×10^3	7	1.2×10^4	6
−7	2.9×10^4	107	9.2×10^4	48

[a]Groups of C57BL/6 mice injected i.p. with FLV on day indicated relative to day of secondary SRBC immunization; mice primed 30 days earlier with SRBC.
[b]Average response for five mice per group 4 days after secondary immunization.

after infection in the C57BL mice as compared to 3 days post infection, indicating that return to normal levels of antibody formation was occurring.

Cellular Immune Response in Mice Susceptible and/or Resistant to FLV

Two parameters of cell-mediated immunity, i.e., macrophage migration inhibition and lymphocyte-mediated cytotoxicity, were examined for the effects of FLV on susceptible and resistant mice. In the migration inhibition (MI) studies BALB/c or C57BL mice were sensitized with Freund's complete adjuvant containing killed mycobacteria. The mice were infected with FLV at various times relative to sensitization with the mycobacteria. Three weeks after sensitization, mouse spleens were removed, and cell suspensions were prepared and placed in capillary tubes. These were incubated in Sykes-Moore chambers in the presence or absence of 50 μg PPD per ml. The areas of migration of cells from the capillary tubes were measured 20 hr later. Inhibition of migration exceeding 20% of control values was considered significant. BALB/c mice infected for 3 days or longer with FLV showed marked inhibition of the MI response, indicating a pronounced suppression of cell-mediated immunity. Conversely, and unlike the effects of FLV infection on humoral immune responses, C57BL mice did not show inhibition of the cellular response by the MI assay at any time after virus infection (Table 3).

Table 3. Macrophage migration inhibition of FLV-infected spleen cells from BALB/c and C57BL/6 mice sensitized with mycobacteria

Time between infection and sensitization[a] (day of FLV infection)	BALB/c		C57BL/6	
	Percent inhibition[b]	Percent of control	Percent inhibition[b]	Percent control
Controls (uninjected)	45	0	40	
−1	22	49	37	98
−3	17	38	40	100
−7	15	33	44	115

[a]Mice sensitized with Freund's complete adjuvant 3 weeks earlier and injected with FLV on indicated day relative to assay day.
[b]Area of migration of spleen cells with PPD/Area of migration of spleen cells without PPD × 100.

Results of measurement of lymphocyte cytotoxicity to allogeneic mastocytoma cells were consistent with those obtained for the humoral immune response (Figure 1, Table 4). BALB/c mice were immunized with the C57BL/6 lymphoma EL4; C57BL/6 mice were injected with the DBA/2 mastocytoma P815. Cytolysis was measured by the release of ^{51}Cr from target cells exactly as described earlier by Brunner and Cerottini (12) and Henney (13). Eleven days after sensitization spleens were removed from the mice and cell suspensions exposed to target cells labeled with chromium. After 6 hr of incubation the amount of cell lysis was assessed. When mice were injected with FLV 6 days after sensitization the BALB/c splenocytes showed significant inhibition of expected cytolysis, whereas responses of C57BL/6 mice were normal (Figure 1). Regardless of the ratio of spleen cells to target cells similar observations were recorded. However, when the level of cytolytic killing of target cells was determined at various times after FLV infection a transient suppression of responsiveness was usually noted (Table 3). This suggests that the presence of virus may indeed have an effect on some cell-mediated immune response in virus resistant mice.

DISCUSSION AND CONCLUSIONS

These studies concerning the relationship of immunologic impairment associated with FLV infection and genetic susceptibility to virus infection per se suggest a correlation between susceptibility to leukemogenesis and immunosuppressive effects of virus replication. C57BL/6

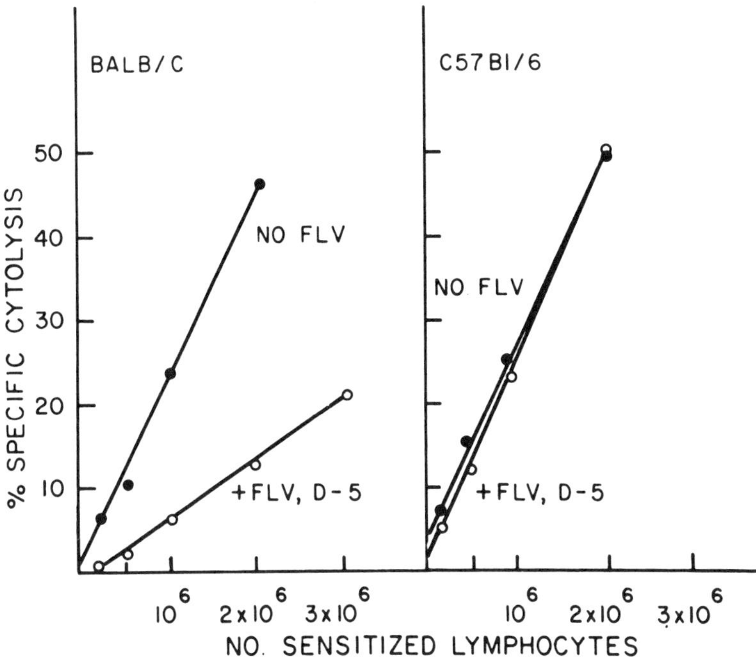

Figure 1. The effect of 500 ID_{50} FLV on lymphocyte mediated cytolysis of allogeneic tumor cells. "Sensitized lymphocytes" refers to splenic lymphocytes from either BALB/c mice sensitized with 3×10^7 C57BL/6 EL4 lymphoma cells or C57BL/6 mice sensitized with 3×10^7 DBA/2 mastocytoma cells. Mice were infected 6 days after sensitization, which was 5 days before assay. The standard deviation was less than 4% at lymphocyte/target cell ratios of 20:1 in noninfected mice.

mice, which are highly resistant to FLV-induced leukemogenesis because of specific genetic control mechanism(s), were found to be highly resistant not only to FLV-induced leukemogenesis but also relatively resistant to immunosuppression. In the assay for leukocyte migration inhibition the virus had no measurable immunosuppressive effects on resistant mice and only a transient depression of immune competence could be detected with other assays (i.e., lymphocyte-mediated cytotoxicity and primary or secondary humoral responses to SRBC). However, it should be noted that the effect of FLV on the secondary humoral antibody response was much more pronounced in terms of 7S as compared to 19S antibody formation, and recovery of hemolysin formation was relatively delayed for IgG producing PFC. Conversely, susceptible BALB/c mice displayed a consistent and progressive suppression of all immune parameters tested in these studies, as had been demonstrated previously in a number of other studies

Table 4. Specific cytolytic activity of lymphocytes for allogeneic target cells at different times after FLV infection[a]

Day of FLV infection relative to time of sensitization	Days post-FLV at assay	Number BALB/c lymphocytes required for 10% specific lysis of targets ($10^4/10^5$)[b]	Ratio FLV infected: non-infected	Number C57BL/6[c] lymphocytes required for 10% specific lysis of targets ($10^4/10^5$)[d]	Ratio FLV-infected: non-infected
+9	2	$7.4 \pm 0.7 \times 10^5$	2.1	$2.9 \pm 1.4 \times 10^5$	1.1
+6	5	$1.6 \pm 0.2 \times 10^6$	4.6	$3.2 \pm 1.2 \times 10^5$	1.2
+3	8	$3.1 \pm 0.8 \times 10^6$	8.9	$3.6 \pm 1.5 \times 10^5$	1.3
0	11	$2.5 \pm 0.4 \times 10^7$	71.0	$2.6 \pm 1.0 \times 10^5$	1.0
−4	15	$3.9 \pm 0.4 \times 10^7$	111.0	$16.1 \pm 8.2 \times 10^5$	6.0
−8	19	1.0×10^8	1,000.0	$4.7 \pm 1.3 \times 10^5$	1.7
−12	23			$3.0 \pm 1.1 \times 10^5$	1.1
Non-infected		$3.5 \pm 1.0 \times 10^5$	1.0	$2.7 \pm 0.9 \times 10^5$	1.0

[a]FLV dose = 500 ID_{50}.
[b]BALB/c mice were sensitized with 3×10^7 C57BL/6 EL4 lymphoma and assayed 11 days later on autologous targets.
[c]C57BL/6 mice were sensitized with 3×10^7 DBA/2 mastocytoma cells and assayed 11 days later on autologous targets.
[d]Plots of the percent specific cytolysis for each mouse were made, and the number of lymphocytes needed to lyse 10% of the targets were determined from the graph; the mean ± S.D. for each group of three to four mice was calculated.

(6,7,14). Thus the immunosuppressive effects seemed to follow the pattern of the presence of virus and its ability to replicate in the mice, since FLV persisted in high titers in BALB/c mice but only transiently detectable virus was present in C57BL/6 mice. Whether the genetic relationship of susceptibility to immunosuppression was under direct genetic control or, alternatively, whether it may reflect a genetic susceptibility to leukemogenesis per se with only an indirect relationship to immune competence, remains to be determined.

Susceptibility or resistance of mice pertaining to immune competence has been associated with the *H-2* genotype in mice; $H\text{-}2^b$ (i.e., C57BL/6 mice) being resistant and $H\text{-}2^d$ (BALB/c mice) being susceptible. However, genetic control may be even more restrictive than this within the *H-2* locus (1,2). For example, Lilly (15) has suggested that control may involve *Ir* genes. However, appropriate studies have not yet been performed that would test this hypothesis. Nevertheless, Lilly has provided a precedent for such a control that seems applicable to the FLV model since his group has demonstrated that a gene controlling resistance to Gross virus infection, designated *RGV-1*, is located at the K region of the H-2 complex (15). Since this is in close proximity to the *Ir* gene that controls some immune responses, gene(s) regulating immunologic reactivity during FLV infection may be similarly located in or near this locus.

Studies utilizing congenic F_1 mice, as performed by Chesebro, Wehrly, and Stimfling (3), to determine sensitivity of recovery from FLV infection would be most useful to ascertain whether there is a concomitant sensitivity to immune responsiveness. Another approach would be to use strains of mice that are SFFV resistant and LLV susceptible or vice versa to determine if either of these components of the FLV complex are directly associated with immunosuppression. To date it has been quite difficult to prepare "pure" virus components in order to study these effects not only on susceptibility to disease in different mouse strains but also in terms of immunosuppression of humoral and/or cellular responses.

It should be noted that there have been inconsistencies in reports from several laboratories concerning the immune competence of resistant strains of mice injected with murine leukemia viruses. Furthermore, even in those susceptible mouse strains that cause both leukemogenesis and immunosuppression after leukemia virus infection, it is not yet known whether B or T cells are preferentially affected or even if macrophages may be the principal target cells. Furthermore, recent studies with completely in vitro model systems suggest that not only purified murine leukemia viruses per se can affect immune responses, but also soluble factors derived from the host infected with the virus may mediate some level of immunosuppression. The genetic control of formation of such "factors'" and their effects on target cells are not known. Nevertheless, it seems likely that it will be possible to determine by further studies whether susceptibility to immunosuppression is under direct genetic control, separate from regulation of susceptibility to leukemogenesis per se, and whether or not genes involved in such control lie within the H-2 region associated with the *FV* alleles or with other as yet unidentified genes.

SUMMARY

The involvement of immune response genes in the control of susceptibility or resistance of mice to the immunosuppressive effects of murine leukemia viruses is under active investigation. Earlier studies in this and in other laboratories indicated that susceptible mice infected with Friend leukemia virus components are markedly depressed in their ability to respond with humoral antibody formation to an antigen such as sheep erythrocytes or to mount an effective cell-mediated immune response to mycobacteria or alloantigens as determined by various in vitro assays. Furthermore, genetically resistant C57BL mice which normally do not support replication of FLV show a transient suppression of antibody formation shortly after infection of FLV. Depressed antibody formation in the resistant mice corresponded to a transient replication of virus in the spleen. The cellular immune response to mycobacteria as determined by the macrophage migration inhibition assay showed marked susceptibility to suppression in FLV susceptible mice. However, in the virus resistant strain, i.e., C57BL/6 mice, no impairment of cellular response to mycobacteria as assayed by the MI assay was evident. The alloantigenic reaction of T lymphocytes against target tumor cells was markedly inhibited in virus susceptible mice. Virus resistant mice showed a transient inhibition of the cytolytic response, but after virus was no longer detectable in the spleen of these mice the cytolytic reaction to target cells was much less inhibited. It seems likely that a relationship exists between susceptibility to virus replication in mouse strains and impairment of immune responses at the humoral or cellular level. However, this relationship may be indirect since there is no evidence at present that immune response genes per se are directly related to those genes that control susceptibility to virus infection in vivo or in vitro. Further studies are warranted to examine the complex interaction between murine virus leukemogenesis and susceptibility or resistance to impaired immune responses.

LITERATURE CITED

1. Lilly, F., and Pincus, T. 1973. Genetic control of murine leukemogenesis. Adv. Cancer Res. 17:231-277.
2. Ceglowski, W. S., Campbell, B. P., and Friedman, H. 1975. Immunosuppression by leukemia viruses. XI. Effect of Friend Leukemia virus on humoral immune competence of leukemia-resistant C57BL/6 mice. J. Immunol. 114:231-236.
3. Chesebro, B., Wehrly, K., and Stimfling, J. 1974. Host genetic control of recovery from Friend leukemia virus-induced splenomegaly. Mapping of a gene within the major histocompatibility complex. J. Exp. Med. 140:1457-1467.
4. Tucker, H. S. G., Weens, J., Tsichlis, P., Schwartz, R. S., Khiroya, R., and Donnelly, J. 1977. Influence of H-2 complex on susceptibility to infection by murine leukemia virus. J. Immunol. 118:1239-1243.
5. Cerny, J., Essex, M., Rich, M. A., and Hardy, W. D., Jr. 1975. Expression of virus associated antigens and immune cell functions during spontaneous regression of the Friend viral murine leukemia. Int. J. Cancer 15:351-365.
6. Dent, P. B. 1972. Immunodepression by oncogenic viruses. Prog. Med. Virol. 14:1-35.
7. Mortensen, R. E., Ceglowski, W. S., and Friedman, H. 1974. Leukemia virus induced immunosuppression. X. Depression of T-cell mediated cytotoxicity after infection of mice with Friend leukemia virus. J. Immunol. 112:2077-2086.
8. Ceglowski, W. S., LaBadie, G. U., Mills, L., and Friedman, H. 1973. Suppression of the humoral immune response by Friend leukemia virus. In W. S. Ceglowski and H. Friedman (eds.), Virus Tumorigenesis and Immunogenesis. pp. 167-177. Academic Press, New York.
9. Ceglowski, W. S., and Friedman, H. 1969. Immunosuppression by leukemia viruses. II. Cytokinetics of appearance of hemolysin-forming cells in infected mice during the anamnestic response to sheep erythrocytes. J. Immunol. 102:338-346.
10. Bloom, B., and Bennet, B. 1966. Mechanism of a reaction in vitro associated with delayed type hypersensitivity. Science 153:80-82.
11. David, J. R. 1966. Delayed hypersensitivity in vitro: Its mediation by cell-free substances formed by lymphoid cell-antigen interactions. Proc. Nat. Acad. Sci. 36:72-77.
12. Brunner, K. T., and Cerottini, J. C. 1971. In vitro assay of target cell lysis by sensitized lymphocytes. In B. Amos (ed.), Progress in Immunology, pp. 385-393. Academic Press, New York.
13. Henney, C. S. 1971. Quantitation of the cell mediated immune responses. The number of cytolytically active mouse lymphoid cells induced by

immunization with allogeneic mastocytoma cells. J. Immunol. 107: 1558-1566.
14. Ceglowski, W. S., LaBadie, G. U., and Mascio, A. A. 1976. Effects of leukemia viruses on cell mediated immune responses. In R. L. Crowell, H. Friedman, and J. E. Prier (eds.), Tumor Virus Infection and Immunity, pp. 165-173. University Park Press, Baltimore.
15. Lilly, F. 1971. The influence of H-2 type on Gross virus leukemogenesis in mice. Transplant. Proc. 3:1239-1241.

Infection, Immunity, and Genetics
Edited by Herman Friedman, T. Juhani Linna, and James E. Prier

ACTIVATION OF ENDOGENOUS RETROVIRUSES AND THE MAJOR HISTOCOMPATIBILITY COMPLEX

A. J. Dennis, A. D. Barker, and J. M. Rice

Genetic information for potentially tumorigenic endogenous Type C RNA viruses (also designated MuLV and more recently retroviruses) is present in the cells of virtually all mouse strains (see (3) for review). These viruses have been classified as ecotropic (N-, B-tropic) or xenotropic (S-tropic) based on their ability to replicate in cells of the same species or those of species other than the species of origin, respectively (13,18). Host cell restriction of N- and B-tropic endogenous viruses is controlled by the *FV-1* gene (21).

Although the cellular control mechanism(s) involved in regulating the expression of most of the endogenous viruses is not presently understood, several modes of retrovirus activation have been identified. Endogenous virus activation has been associated with the aging process both in vivo and in vitro (19), clonal variation in vitro (20,32), cellular differentiation (4), in vitro transformation (32), and tumorigenesis in certain mouse strains with a high incidence of spontaneous neoplastic disease (e.g., AKR, NZB, SJL) (12).

In addition to these examples of "spontaneous" activation, retroviruses have been activated by halogenated pyrimidines (1,23), inhibitors of protein synthesis (2), and certain lymphocyte mitogens (11). Last, and perhaps most relevant to the problem of tumorigenesis,

Supported by Battelle Institute Grant B1333-1200.

activation of retroviruses has been associated with graft rejection (15), the graft versus host reaction (GVHR) (16), and the mixed lymphocyte reaction (MLR) (5,16). It also has been demonstrated that some retroviruses replicate preferentially in lymphocytes (31), and most immunologic activation studies performed to date have involved interactions between histoincompatible lymphocytes.

Although the mechanism(s) of immunologic activation of endogenous viruses is complex and remains unresolved, the phenomenon has been repeatedly associated with differences in the major histocompatibility complex (MHC). Antigenic differences in the MHC can initiate specific immunologic recognition and subsequent proliferation of lymphocyte subpopulations.

The recognitive phase of this sequence of immunologic events may result from either minor or major differences in the H-2 region (6) and in the case of major histoincompatibility correlates with generation of Killer cells (2) and may be associated with virus activation (6). Non-specific lymphocyte blastogenesis alone is apparently insufficient to activate retroviruses (26).

The retroviruses activated as a consequence of immune functions have been linked both directly and indirectly to increased tumorigenesis in host animals. Hirsch et al. have observed activation of retroviruses following skin transplantion across major H-2 barriers in immunosuppressed mice that retained their grafts (15). These investigators suggest that immunologically activated retroviruses may be responsible for the increased incidence of lymphoreticular neoplasms seen in immunosuppressed kidney transplant patients. Induction of the GVHR in F_1 mice using parental lymphocytes also results in the production of virus and subsequently the development of lymphoreticular tumors (14,16). The etiological role of these viruses in tumorigenesis has been further established by the induction of lymphoreticular tumors in newborn syngeneic recipients following inoculation of cell-free extracts prepared from the tissues of mice undergoing GVHR (5).

An increasing body of evidence indicates that the lymphoid system, activation of retroviruses, and neoplastic disease states are intimately related. However, the significance of this relationship in the normal host, the mechanism of immune activation of retroviruses, and the relationship of these problems to the MHC are all questions that remain to be solved. The studies reported here were designed to investigate the relationship between activation of retroviruses across minor and major H-2 barriers, tropisms of activated viruses, and generation of cytotoxic lymphocytes.

EXPERIMENTAL DATA

The cell line used in the studies reported here was derived from non-virus-producing BALB/c 3T3 C1-A31 cells. These cells were morphologically transformed at high passage by culturing at super confluency. The cloned, virus-negative transformed variant, designated HP-3T3, was found to produce tumors in syngeneic BALB/c mice. The BALB/3T3 C1-A31 and HP cell lines were tested for inducibility of endogenous virus by exposure of subconfluent monolayers to bromodeoxy uridine (BrdU). As can be seen in Table 1, both lines were induced to produce high levels of endogenous virus.

A second cell line designated S (spontaneous for B-tropic virus production) was derived by passaging the HP cell line twice in BALB/c mice and subculturing the resultant tumors. The system has recently been presented in detail (7).

Ficoll-Hypaque-separated lymphocytes from the thymuses, spleens, and lymph nodes of 8-week-old BALB/c, DBA/2, or C57BL/6 mice (Jackson Laboratories) were co-cultivated for 3 days with HP cells (60% confluent) and the lymphocytes assessed for cytotoxicity to HP target cells (25:1 effector:target cells) using a ^{51}Cr release assay. The monolayers were subcultured and assayed at regular intervals for reverse transcriptase activity and the host range restrictions of the activated retroviruses were determined.

Table 1. Reverse transcriptase activity in BrDU-induced normal and spontaneously transformed BALB/c 3T3 clone A31 cells

Cell line	In vitro passage number	BrDU (32 μg/ml)	Cpm × 10^{-3} [a]
3T3 A31	4	−	1.5
	4	+	91.6
	13	−	2.8
	13	+	158.3
HP-3T3 [b]	20	−	0.7
	20	+	105.9
	40	−	0.4
	40	+	250.0

[a] Counts per minute are for 15 ml of cell culture medium from approximately 1×10^7 cells (90–100% confluency). Cpm are presented minus control tube counts which varied from 2.0 to 12.0 × 10^3 cpm.

[b] HP-3T3 is the designation for a "high passage" cloned transformant of the parent 3T3 C1-A31 line.

The reverse transcriptase (RT) assay routinely performed on test supernatants was modified from that reported by Ross et al. (27). The supernatants from co-cultivated cultures were always assayed fresh to maximize recovery of virus RT activity. RT values in experimental cultures had to be at least 10,000 counts above control values to be considered RT positive. Background counts ranged from 2,000 to 12,000 cpm in these studies. The host range restrictions of activated viruses (N, B, and/or [S] tropism) were determined on NIH/3T3 cells, BALB/3T3 cells, and SIRC cells, respectively. Supernatants from these cultures were assayed for RT activity at each subsequent subculture.

As shown in Table 2 endogenous viruses with different host range restrictions were chemically induced from both the parental BALB/c 3T3 cells and the HP variant line. Following BrdU induction, the normal 3T3 C1-A31 cells produced high levels of B-tropic virus while the transformed HP cells demonstrated moderate levels of B-tropic virus and high levels of xenotropic virus. Both cell lines consistently showed very low titers of positive activity on the NIH indicator cell line. Activity did not remain positive with successive subculture (probably indicating the subversion of host range restriction by high multiplicities of infection). In certain instances the XC plaque assay (17) was employed to confirm viral RT activity. Cultures were considered positive for ecotropic virus production only if they were clearly positive in both assays.

Table 2. Host range restriction of chemically induced endogenous viruses

	Ecotropic[a]		
Cell line	NIH/3T3 (N-tropic)	BALB/3T3 (B-tropic)	SIRC (xenotropic)
3T3 C1-A31 (BrdU induced)	±	+ + +[b]	
HP (BrdU induced)	±	+ + +	+ + + +

[a]The tropisms of viruses were determined by seeding 1.0 × 10^6 BALB/c 3T3 C1-A31, NIH/3T3, or SIRC (normal rabbit corneal cells) in 250 ml plastic tissue culture flasks in complete medium containing 2.0 μg/ml polybrene 24 hr before infection. Quadruplicate cultures of each cell type then were exposed to 1.5 ml of undiluted, filtered (0.45 μ millipore) or purified virus for 1 hr. The cells were subcultured at a split ratio of 1:20 every 7 days for three passages. Four days following each subculture, the supernatants were decanted and stored at − 70 °C for both RT and XC assay.

[b]Values indicate relative quantities of virus produced as measured by XC. Results are shown as positive only if they were positive for both RT and XC tests.

To examine the effects of minor or major differences in the MHC on *in vitro* immunologic activation of endogenous viruses, 3T3 C1-A31 and HP cells were co-cultivated with lymphocytes, as previously described. Thymuses, spleens, and lymph nodes from 8-week-old normal BALB/c (*H-2d*), DBA/2 (*H-2d*), and C57BL/6 (*H-2b*) mice were used as sourced of lymphocytes. BALB/c mice were syngeneic with the 3T3 C1-A31 line, DBA/2 mice possessed the same major H-2 specificities with some minor differences, and C57BL/6 animals were histoincompatible with BALB/c mice at a major locus. Ficoll-Hypaque-purified lymphocytes from different lymphoid compartments were employed in the studies; work in our laboratories and in others (10,31) suggests that lymphoid subpopulations may differentially replicate and/or activate retroviruses.

Although all co-cultivation experiments were performed using both 3T3 C1-A31 and HP cell cultures, we observed no detectable retrovirus in any of the experiments using 3T3 C1-A31 cells. Therefore, the negative data obtained from the 3T3 C1-A31 lymphocyte co-cultivations are not presented.

Thymocytes co-cultivated with either 3T3 C1-A31 or HP cells never resulted in activation of detectable levels of endogenous virus, with the exception of C57BL/6 thymocytes co-cultivated with HP cells. A representative series of RT values from these studies is presented in Table 3. The C57BL/6-induced activity was detected after four subcultures, and it remained positive in subsequent passages. In an attempt to identify the source of activated endogenous virus, all lymphocyte HP co-cultivations that resulted in detectable endogenous virus production were reproduced using x-irradiated lymphocytes. As shown in Table 3, x-irradiated thymocyte HP co-cultivation did not result in virus activation. HP cell controls and lymphocyte cultures from the three mouse strains tested were consistently negative for spontaneous virus production in all studies.

Co-cultivation of splenic lymphocytes from both DBA/2 and C57BL/6 mice resulted in high levels of endogenous virus, as measured by RT (Table 4). The quantity of virus activated in the C57BL/6-HP co-cultivation was markedly greater and appeared in earlier passages than the levels seen following co-cultivation with DBA/2 splenocytes. This increased level of endogenous virus production is consistently associated with increased histoincompatibility between the BALB/c-derived HP cell and the lymphocytes used for co-cultivation. X-irradiated DBA/2 or C57BL/6 splenic lymphocytes did not activate endogenous virus. Relative quantitation of RT is based on the use of 15 ml of medium from identical numbers of producer cells.

Table 3. Reverse transcriptase activity from cultures of normal and x-irradiated thymocytes co-cultivated with HP-3T3 cells

Lymphocyte donor	Subculture number post co-cultivation[a]	Cpm × 10^{-3}
BALB/c	P-3	6.4
(*H-2*^{d})	P-4	Neg.[b]
	P-3 (x-irradiated)	N.D.[c]
	P-4 (x-irradiated)	N.D.
	Lymphocytes only	5.6
	HP control P-4	5.6
DBA/2	P-3	Neg.
(*H-2*^{d})	P-4	Neg.
	P-3 (x-irradiated)	N.D.
	P-4 (x-irradiated)	N.D.
	Lymphocytes only	5.9
	HP control P-4	5.6
C57BL/6	P-3	2.9
(*H-2*^{b})	P-4	74.3
	P-3 (x-irradiated)	2.0
	P-4 (x-irradiated)	9.5
	Lymphocytes only	3.0
	HP control P-4	5.6

[a] All 3T3 C1-A31 co-cultivations were negative for endogenous virus production.
[b] Neg. = control cpm higher than experimental cpm.
[c] N.D. = not done.

This differential pattern of activation was also observed when lymph node lymphocytes were employed for co-cultivation. A representative series of experiments are presented in Table 5. There was more virus as detected by the RT assay in the C57BL/6-HP cultures as compared with DBA/2-HP cultures. However, the levels of endogenous virus produced following co-cultivation of HP cells with DBA/2 lymph node lymphocytes was significantly increased over those noted when DBA/2 splenic lymphocytes were employed. As observed in previous experiments, x-irradiated lymph node lymphocytes did not activate endogenous virus when co-cultured with HP cells.

The host range restrictions of the endogenous retroviruses activated in these co-cultivation studies are shown in Table 6. The endogenous retroviruses recovered following co-cultivation of either DBA/2 splenic or lymph node lymphocytes with HP cells was N-tropic. Conversely, endogenous viruses activated following co-cultiva-

Table 4. Reverse transcriptase activity from cultures of normal and x-irradiated splenocytes co-cultivated with HP-3T3 cells

Lymphocyte donor	Subculture number post co-cultivation[a]	Cpm × 10^{-3}
BALB/c	P-3	0.2
(*H-2^d*)	P-4	Neg.[b]
	P-3 (x-irradiated)	N.D.[c]
	P-4 (x-irradiated)	N.D.
	Lymphocytes only	2.3
	HP control P-4	5.6
DBA/2	P-3	43.6
(*H-2^d*)	P-4	83.5
	P-3 (x-irradiated)	4.4
	P-4 (x-irradiated)	3.5
	Lymphocytes only	Neg.
	HP control	5.6
C57BL/6	P-3	325.6
(*H-2^b*)	P-4	414.0
	P-3 (x-irradiated)	1.9
	P-4 (x-irradiated)	2.0
	Lymphocytes only	3.7
	HP control	5.6

[a] All 3T3 C1-A31 co-cultivations were negative for virus production.
[b] Neg. = control cpm higher than experimental cpm.
[c] N.D. = not done.

tion of either C57BL/6 splenic or lymph node lymphocytes with HP cells was shown to be B-tropic.

The direct relationship of histocompatibility differences between lymphoid subpopulations used for co-cultivation and the HP-3T3 cells to endogenous virus activation was further seen in the cytotoxic responses of these in vitro sensitized lymphocytes for labeled HP target cells (Table 7). The highest levels of cytotoxicity were seen in C57BL/6 lymph node and splenic lymphocytes with significant responses noted in lymphocytes from DBA/2 co-cultivations. Lymphocytes from BALB/c co-cultivation experiments did not demonstrate significant cytotoxic responses to HP-3T3 target cells.

DISCUSSION

The data presented demonstrate several significant differences in both quantity and host range restrictions of retroviruses activated

Table 5. Reverse transcriptase activity from cultures of normal and x-irradiated lymph node lymphocytes co-cultivated with HP-3T3 cells

Lymphocyte donor	Subculture number post co-cultivation[a]	Cpm × 10^{-3}
BALB/c	P-3	Neg.[b]
(*H-2[d]*)	P-4	Neg.
	P-3 (x-irradiated)	N.D.[c]
	P-4 (x-irradiated)	N.D.
	Lymphocytes only	Neg.
	HP control P-4	5.6
DBA/2	P-3	97.9
(*H-2[d]*)	P-4	227.0
	P-3 (x-irradiated)	Neg.
	P-4 (x-irradiated)	N.D.
	Lymphocytes only	3.4
	HP control P-4	5.6
C57BL/6	P-3	328.9
(*H-2[b]*)	P-4	363.4
	P-3 (x-irradiated)	1.7
	P-4 (x-irradiated)	7.0
	Lymphocytes only	3.1
	HP control P-4	5.6

[a] All 3T3 C1-A31 co-cultivations were negative for endogenous virus production.
[b] Neg. = control cpm higher than experimental cpm.
[c] N.D. = not done.

Table 6. Host range restriction of viruses activated following co-cultivation of lymphocytes with HP cells

Lymphocyte donor (organ)[a]	Reverse transcriptase activity[b] (cpm × 10^{-3})		
	NIH/3T3	BALB/3T3	SIRC
DBA/2 (spleen)	112.7	Neg.[c]	Neg.
DBA/2 (lymph node)	263.0	11.1	Neg.
C57BL/6 (spleen)	7.7	40.5	1.9
C57BL/6 (lymph node)	7.9	77.7	Neg.

[a] Virus from supernatants of P-4 subculture post co-cultivation.
[b] Indicator cells were assayed at P-2 following infection.
[c] Neg. = control cpm higher than experimental cpm.

Table 7. Cytotoxic response of lymphocytes following co-cultivation with HP-3T3 cells

Lymphocyte donor	Percent cytotoxicity[a]
BALC/c spleen	9.8 ± 3.3[b]
BALB/c lymph nodes	3.4 ± 1.5
DBA/2 spleen	14.7 ± 3.5
DBA/2 lymph nodes	14.5 ± 0.4
C57BL/6 spleen	23.6 ± 1.7
C57BL/6 lymph nodes	38.9 ± 23.1

[a]The assay was performed at an effector:target cell ratio of 25:1. Percent cytotoxicity = cpm exp − spontaneous release/Cpm 100% release − spontaneous release × 100.

[b]Standard error.

either chemically or immunologically from BALB/c 3T3 C1-A31 cells and from the transformed variant of this line (HP cells). Specifically the studies reported here show that:

1. Both the normal cells and the transformed variant could be activated to produce retroviruses by treatment with BrdU
2. B-tropic virus was chemically activated from normal BALB/c 3T3 C1-A31 cells, while both B- and S-tropic viruses were induced from HP cells
3. Endogenous retrovirus was not activated when BALB/c 3T3 C1-A31 cells were co-cultivated with either histocompatible or allogeneic lymphocytes
4. N-tropic retrovirus was activated following co-cultivation of either DBA/2 splenic or lymph node cells with HP cells
5. B-tropic retrovirus was activated in C57BL/6 thymus, spleen, and lymph node—HP co-cultivation
6. The highest level of endogenous retrovirus was activated when lymph node cells were used for co-cultivation
7. The cytotoxic potential of lymphocytes from co-cultivation experiments correlated directly with retrovirus activation
8. Retrovirus was not activated by co-cultivation of HP cells with x-irradiated lymphocytes

The normal and transformed variant (HP) cell lines used in our studies were originally derived from BALB/c mice. This is a low leukemia strain that begins to express endogenous virus (primarily B-tropic) only with increasing age (see 28 for review). Although retrovirus production in this strain is apparently under stringent control(s), both ecotropic (N and B) and xenotropic (S) retroviruses have been chemically activated from BALB/c embryo cells in vitro. It also has been reported that nonvirus-producing transformed variants of BALB/c 3T3 cells contain greater amounts of viral RNA than the parent line and that control of viral expression is altered (32).

Lieber, Sherr, and Todaro (19) have reported that co-cultivation of spleen cells from the mouse strains employed in our studies (among several others studied) with either BALB/c 3T3, NIH 3T3, or SIRC cells resulted in B-tropic virus (BALB/c) or S-tropic virus production (DBA/2 and C57BL/6). These authors noted a change in the host range restriction pattern of the viruses produced when aged BALB/c or C57BL/6 mice were tested.

It was concluded from these experiments that the "target" cell lines were replicating retrovirus that was passively released from the splenic lymphocytes and was not caused by any of the modes of immunologic activation observed by other investigators. The controls for these experiments were freeze-thaw lysates prepared from splenic lymphocytes.

The data from our studies strongly suggest that retroviruses were activated from transformed HP cells via an immune mechanism. This conclusion seems to be warranted in view of the H-2 differences between certain of the lymphocytes used for co-cultivation and the HP cells. This would certainly trigger immune recognition and generation of killer cells (as we have shown). This recognition process has been shown to involve recognition of antigenic differences in the K, D, and Ir regions of the MHC (6).

The following experimental observations further support the immune activation versus passive infection hypotheses: the quantity of retrovirus activated was directly related to the degree of histoincompatibility between test lymphocytes and HP cells and with lymphocyte-mediated cytotoxicity, and the lymphocyte preparation containing the largest percentage of highly differentiated T cells (lymph nodes) consistently activated the greatest quantities of retrovirus from HP cells. Moreover, x-irradiated lymphocytes did not activate retrovirus from HP cells. This latter observation could of course be a consequence of total inactivation of any incorporated proviral genes, but this must be demonstrated experimentally.

Melief et al. (24) also have observed induction of N-tropic virus when BALB/c lymphocytes were co-cultivated with DBA/2 lymphocytes, but killer cells were not generated in these experiments. Co-cultivation of DBA/2 splenic and lymph node lymphocytes with HP cells in our studies also resulted in N-tropic virus production, and a significant level of lymphocyte cytotoxicity against HP cells was observed. The differences in killer cell generation between these two studies may be related to the presence of tumor-associated antigens and/or viral antigens on the surface of transformed HP cells as well as differences in the nature of immunologic recognition between MLC and lymphocyte fibroblast interactions.

Studies by several groups of investigators have demonstrated that retroviruses are activated during the GVHR (see 14 for review). This uncontrolled proliferation of lymphoid cells and retrovirus activation (especially N-tropic virus) have been shown to produce lymphomas (14,24), and also have been implicated in certain autoimmune disorders (14).

Based on these types of studies it could be postulated that *in vivo* immunologic activation of retroviruses (possibly B-tropic viruses) from cells other than those of the lymphoid series may be involved in solid tumor development. Armstrong et al. recently reported lymphoreticular tumor development following inoculation of an immunologically activated B-tropic retrovirus (5).

Although we can only speculate as to the mechanism(s) of control of retrovirus production, it has been shown in at least two instances, FV-1 related host range restriction (21) and AKV-1 control of MuLv expression in AKR mice (29), that virus maturation and occurrence of leukemia, respectively, are closely associated with the H-2 haplotype (9,22).

That the MHC is closely linked to control of retrovirus production is further suggested by the activation of these viruses by immunologic mechanism(s), the presence of anti-MuLv antibodies in H-2 alloantisera (25), and the selective incorporation of H-2 antigens during the budding of Friend virus from infected cells (9). These latter observations suggest that H-2 may somehow effect control over some final phase of retrovirus replication, and/or be involved in the release of virus at specific sites on the cell membrane.

Questions concerning regulation of retroviruses and their activation require systems for study that include cells in various phases of retrovirus production and/or expression, which are oncogenic in vivo. Todaro (32) has suggested that spontaneously transformed cell lines may permit detection of new viral RNA and specific viral proteins in the absence of virus production. This would facilitate understanding the mechanism(s) by which retrovirus genes are controlled and their relationship to neoplastic disease. We feel that the system described here is potentially useful in experimentally approaching certain of these questions.

ACKNOWLEDGMENT

We thank Ms. Karen Huston for her excellent assistance in the preparation of this manuscript.

LITERATURE CITED

1. Aaronson, S. A. 1971. Chemical activation of focus forming virus from nonproducer cells transformed by murine sarcoma virus. Proc. Nat. Acad. Sci., U.S.A. 68:3069–3072.
2. Aaronson, S. A., and Dunn, C. Y. 1974. High frequency C type induction of inhibitors of protein synthesis. Science 183:422–424.
3. Aaronson, S. A., and Stephenson, J. R. 1976. Endogenous type C RNA viruses of mammalian cells. Biochem. Biophys. Acta 458:323–354.
4. Aoki, T. 1974. Murine type C RNA viruses: A proposed reclassification, other possible pathogenicities, and a new immunologic function. J. Nat. Cancer Inst. 52:1029–1034.
5. Armstrong, M. Y. K., Ruddle, N. H., Lipman, M. B., and Richards, F. K. 1973. Tumor induction by immunologically activated murine leukemia virus. J. Exp. Med. 137:1163–1179.
6. Bach, F. H., Widmer, M. B., Bach, M. I., and Klein, J. 1972. Serologically defined and lymphocyte-defined components of the major histocompatibility complex in the mouse. J. Exp. Med. 136:1430–1438.
7. Barker, A. D., Dennis, A. J., and Moore, V. S. 1977. Immunologic effects of endogenous virus-producing tumors in thymectomized or splenectomized BALB/c mice. Fed. Proc. 36:1291.
8. Benveniste, R. E., Todaro, G. J., Scolnick, E. M., and Parks, W. P. 1973. Partial transcription of murine type C viral genomes in BALB/c cells lines. J. Virol. 12:711–720.
9. Bubbers, J. E., and Lilly, F. 1977. Selective incorporation of H-2 antigenic determinants into Friend virus particles. Nature 266:458–459.
10. Dennis, A. J., and Axler, D. A. 1974. *In vitro* immune activation of endogenous "C" type virus. J. Retic. Soc. (Abs. suppl.) 16:45.
11. Greenberger, J. S., Phillips, S. M., Stephenson, J. R., and Aaronson, S. A. 1975. Induction of mouse type C RNA virus by lipopolysaccharides and Concanavalin A. J. Immunol. 115:317–320.
12. Gross, L. 1970. Oncogenic Viruses. Pergamon Press, Oxford.
13. Hartley, J. W., Rowe, W. P., and Huebner, R. J. 1970. Host range restrictions of murine leukemia viruses in mouse embryo cell cultures. J. Virol. 5:221–225.
14. Hirsch, M. S. 1976. Immune activation of endogenous viruses. In R. L. Crowell, H. Friedman, and J. E. Prier (eds.), Tumor Virus Infections and Immunity, pp 175–186. University Park Press, Baltimore.
15. Hirsch, M. S., Ellis, D. A., Kelly, A. P., Proffitt, M. R., Black, P. H., Monaco, A. P., and Wood, M. L. 1957. Activation of C type viruses during skin graft rejection in the mouse. Interrelationships between immunostimulation and immunosuppression. Int. J. Cancer 15:493–502.

16. Hirsch, M. S., Phillips, S. M., Solnick, C., Black, P. H., Schwartz, R. S., and Carpenter, C. B. 1972. Activation of leukemia viruses by graft-versus-host and mixed lymphocyte reactions *in vitro.* Proc. Natl. Acad. Sci. 69:1069–1072.
17. Klement, V., Rowe, W. P., Hartley, J. W., and Pugh, W. E. 1969. Mixed culture cytopathogenicity: A new test for growth of murine leukemia viruses in tissue culture. Proc. Natl. Acad. Sci. 63:753–758.
18. Levy, J. A. 1970. Demonstration of biologic activity of a murine leukemia virus of New Zealand black mice. Science 170:326–327.
19. Lieber, M. M., Sherr, C. J., and Todaro, G. J. 1974. S-tropic murine type C viruses: Frequency of isolation from continuous cell lines, leukemia virus preparations, and normal spleens. Int. J. Cancer 13:587–598.
20. Lieber, M. M., and Todaro, G. 1973. Spontaneous and induced production of endogenous type C RNA virus from a clonal line of spontaneously transformed BALB/3T3. Int. J. Cancer 11:616–627.
21. Lilly, F. 1970. FV-2: Identification and location of a second gene governing the spleen focus response to Friend leukemia virus in mice. J. Natl. Cancer Inst. 45:163–169.
22. Lilly, F., Duran-Reynals, M. L., and Rowe, W. P. 1975. Correlation of early murine leukemia virus titer and H-2 type with spontaneous leukemia in mice of the BALB/c × AKR cross: A genetic analysis. J. Exp. Med. 141:882–889.
23. Lowry, D. R., Rowe, W. P., Teich, N., and Hartley, J. W. 1971. Murine leukemia viruses: High frequency activation *in vitro* by 5-iododeoxy uridine and 5 bromodeoxy uridine. Science 174:155–166.
24. Melief, C. J. M., Syamal, D., Stephen, L., and Schwartz, R. S. 1974. Immunologic activation of murine leukemia viruses. Cancer 34:1481–1487.
25. Milner, R. J., Henning, R., and Edelman, G. M. 1976. Identification of the murine leukemia viral antigens detected on mouse cells by H-2 alloantisera. Eur. J. Immunol. 6:603–607.
26. Phillips, M. S., Stephenson, J. R., Greenberger, J. S., Lane, P. T., and Aaronson, S. A. 1976. Release of xenotropic type C RNA virus in response to lipopolysaccharide: Activity of lipid-A-portion upon B lymphocytes. J. Immunol. 116:1123–1128.
27. Ross, J., Scolnick, E. M., Todaro, G. J., and Aaronson, S. A. 1971. Separation of murine cellular and murine leukemia virus DNA polymerase. Nature New Biol. 231:163–167.
28. Rowe, W. P. 1972. Genetic factors in the natural history of murine leukemia virus infection: GHA Clowes memorial lecture. Cancer Res. 33:3061–3068.
29. Rowe, W. P. 1972. Studies of genetic transmission of murine leukemia virus by AKR mice. I. Crosses with FV-1^n strains of mice. J. Exp. Med. 136:1272–1285.
30. Rowe, W. P., and Hartley, J. W. 1972. Studies of genetic transmission of murine leukemia virus by AKR mice. II. Crosses with FV-1^b strains of mice. J. Exp. Med. 136:1286–1301.
31. Ruddle, N. H., Armstrong, M. Y. K., and Richards, F. 1976. Replica-

tion of murine leukemia virus in bone marrow derived lymphocytes. Proc. Nat. Acad. Sci. U.S.A. 73:3714-3718.

32. Todaro, G. 1972. Spontaneous release of type C virus from clonal lines of "spontaneously" transformed BALB/3T3 cells. Nat. New Biol. 240:157-160.

Infection, Immunity, and Genetics
Edited by Herman Friedman, T. Juhani Linna, and James E. Prier

AUTOIMMUNITY, INFECTIONS, AND GENETICS

T. J. Linna

In the very beginning of this century, Ehrlich and Morgenroth coined the term "horror autotoxicus" to express the absence of manifest autoreactivity in the normal individual (1). Much later, Burnet's formulation of the clonal selection theory (2) provided a theoretical base for understanding the immunological discrimination between self and nonself. The clonal selection theory has influenced much of the thinking in immunology in the past years, particularly our interpretations of autoimmunity. The recent findings that autoreactivity is suppressed rather than absent in the normal individual (3, 4) have provided evidence that the clonal selection theory may be oversimplified, and that manifest autoimmunity may be caused by a failure in the delicate network of checks and balances of the immune system. A broken T cell tolerance (5, 6) is one currently attractive way to explain the breakdown of self-tolerance.

Much progress has occurred in recent years and is still taking place, which makes it important and interesting to focus on the relationship between autoimmunity, infections, and genetics. Full coverage of this area would require its own symposium and progress volume. The following chapters highlight some of the recent developments in crucial parts of this field.

Williams and Yunis (7) discuss the influence of genetic factors on disease, particularly in man. The association between the major histocompatibility complex (MHC) and many disease entities is getting more impressive by the day. Associations of varying strength exist between the MHC and diseases with pathogenetic immunological components, such as malignancies, infectious diseases, and diseases with autoaggressive features. The common denominator for all these diseases may well be the immunoregulatory functions of the MHC. This arca will certainly provide much useful information, not only for disease classification and genetics, but for our understanding of the genetic regulation of the immune responses, and the interplay in health and disease between the immune system, the host, and its invaders.

Progress has been made in recent years toward the understanding of the causes of multiple sclerosis, a well-known disease entity. However, many key questions remain unanswered. This disease exemplifies steps taken toward the understanding of the genetic, immunological, microbiological, and epidemiological aspects of the pathogenesis of diseases with autoaggressive features, but also the challenge remaining for us to fully understand the pathogenesis of the disease, and to ultimately develop a cure. Williams and Yunis briefly discuss genetic aspects of multiple sclerosis, and Silberberg (8) brings us up to date on the progress made toward the understanding of the etiology of this probably multifactorial disease.

Experimental models strongly implicating the importance of genetics in autoimmunity have been available for a long time. Two such classic models are discussed here, namely the obese (OS) chicken (9) and the New Zealand mouse (10). Kite informs us of the latest developments with regard to the influence of the MHC on thyroid disease expression in the OS animals (11). In both these syndromes, a microbial (virus) infection has been proposed in the pathogenetic process, but it is presently unclear how central a role a virus would play in etiology. It is clear that there is a derangement of immunological responsiveness in both diseases. An inadequately functioning suppressor (T) cell system has been proposed to have a central role in these and other autoimmune diseases. Although the final relevance of suppressor cells for autoimmunity remains to be determined, Kite's contribution underlines the complexity of the influences of the immune system on the autoimmune process. It is highly probable, from the data available in the OS chicken model, that the bursa-dependent, antibody-forming system is instrumental in the initiation of cellular injury. These findings agree well with the "surveillance" role of bursa-dependent immunity (12) and antibodies (13) in the sister discipline tumor immunology.

Ann Gabrielsen proposed a few years ago a novel way to look at the immune complex-induced damage in New Zealand mice and perhaps also in systemic lupus erythematosus, namely that the basic defect would be an inappropriate handling of nuclear material, made available from the final maturation steps of cells of the erythropoietic series. This would result in immune complex formation in antigen excess and immune complex deposition, particulaly in the kidney glomeruli (14). The experimental evidence from work in her laboratory with the (NZB × NZW)F_1 mice supporting the hypothesis is now quite substantial. The erythrosuppressive therapy results are impressive, and the implications for pathogenesis and perhaps even therapy of SLE are close at hand (15).

At present, it appears that autoimmune diseases and phenomena are of multifactorial genesis, with microbiological and genetic components playing major roles. These diseases include not only infrequent "experiments of nature" of interest to the immunologist, but also common diseases such as juvenile onset diabetes (7, 16). An understanding of the multifaceted problems of autoimmunity will certainly help to understand the normal functions of the immune system, and also the relationship between immunity and malignancy and host-parasite relationships in general.

LITERATURE CITED

1. Ehrlich, P., and J. Morgenroth. 1901. Über Hämolysine. Berl. Klin. Wochenschr. 38:251-257.
2. Burnet, F. M. 1959. The Clonal Selection Theory of Acquired Immunity. University Press, Cambridge.
3. Cohen, I. R., and H. Wekerle. 1972. Autosensitization of lymphocytes against thymus reticulum cells. Science 176:1324-1325.
4. Bankhurst, A. D., G. Torrigiani, and A. C. Allison. 1973. Lymphocytes binding human thyroglobulin in healthy people and its relevance to tolerance for autoantigens. Lancet 1:226-230.
5. Chiller, J. M., G. S. Habicht, and W. O. Weigle. 1970. Cellular sites of immunologic unresponsiveness. Proc. Nat. Acad. Sci. USA 65:551-556.
6. Gershon, R. K., and K. Kondo. 1972. Tolerance to sheep red cells: Breakage with thymocytes and horse red cells. Science 175:996-997.
7. Williams, R. M., and E. J. Yunis. 1978. Genetics of human immunity and its relation to disease. In H. Friedman, T. J. Linna, and J. E. Prier (eds.), Infection, Immunity, and Genetics, p. 121. University Park Press, Baltimore.
8. Silberberg, D. H. 1978. Multiple sclerosis, the HLA system, and viruses. In H. Friedman, T. J. Linna, and J. E. Prier (eds.), Infection, Immunity, and Genetics, p. 000. University Park Press, Baltimore.
9. Cole, R. K., J. H. Kite, and E. Witebsky. 1968. Hereditary autoimmune thyroiditis in the fowl. Science 160:1357-1358.
10. Howie, J. B., and L. O. Simpson. 1978. The immunopathology of the NZB mice and their hybrids. In P. A. Miescher and J. H. Muller-Eberhard (eds.), Textbook of Immunopathology, pp. 247-278. Grune & Stratton, New York.
11. Kite, J. H. 1978. Genetic control of thyroiditis. In H. Friedman, T. J. Linna, and J. E. Prier (eds.), Infection, Immunity, and Genetics, p. 157. University Park Press, Baltimore.

12. Linna, T. J., C. Hu, and K. M. Lam. 1976. Host protection by the antibody-forming system in malignancy. In M. Feldman and A. Globerson (eds.), Immune Reactivity of Lymphocytes, pp. 415-421. Plenum, New York.
13. Hu, C., and T. J. Linna. 1976. Serotherapy of avian reticuloendotheliosis virus-induced tumors. Ann. New York Acad. Sci. 277:634-646.
14. Gabrielsen, A. E. 1974. Lupus erythematosus: A disease of DNA discard? Lancet II, 1116-1118.
15. Gabrielsen, A. E. 1978. Murine lupus erythematosus and its clinical implications. In H. Friedman, T. J. Linna, and J. E. Prier (eds.), Infection, Immunity, and Genetics, p. 147. University Park Press, Baltimore.
16. Irvine, W. J. 1977. Autoimmunity in endocrine diseases. In T. E. Mandel, C. Cheers, C. S. Hosking, I. F. C. McKenzie, and G. J. V. Nossal (eds.), Progress in Immunology III. Australian Academy of Science, Canberra City.

Infection, Immunity, and Genetics
Edited by Herman Friedman, T. Juhani Linna,
and James E. Prier

GENETICS OF HUMAN IMMUNITY AND ITS RELATION TO DISEASE

R. Michael Williams and Edmond J. Yunis

The major histocompatibility complex (MHC) of all studied mammals contains genes that are responsible for the recognition of individual differences by transplantation within a species. In addition to the genes that code for the molecules that act as targets for homograft rejection, there are numerous other genes associated with this complex that influence host immunity.

The MHC of man—HLA—is located on the sixth chromosome occupying a region of less than two recombination units. Based particularly on studies of the MHC of the mouse—H-2—this system has come to represent a complex series of interacting genes with common evolutionary histories and rapidly unfolding interdependent functions in the maintenance of host integrity.

A simplified map of the sixth chromosome with some of the best identified marker genes is included in Figure 1. Approximate genetic distances in recombination units (Centimorgans) is indicated by the boldface numbers between the tic marks. For a more detailed summary of this subject, readers should consult the many recent reviews of this subject (1), including one with relatively exhaustive technical detail as utilized in our laboratory (2).

Among the most intriguing genes in this complex are the so-called immune response (*Ir*) genes, which have been clearly shown to be distinct members of the MHC, but for which as yet no definitive gene product has been identified. Studies in numerous experimental animals have demonstrated that these immune response genes regulate host reactivity to various types of antigenic specificities, including

Research performed in the authors' laboratories was supported by: NIH Grants Nos. CA 19791 (RMW) and CA 19589-02, CA 20531-01 (EJY) and Grant No. IM-149 from the American Cancer Society (RMW).

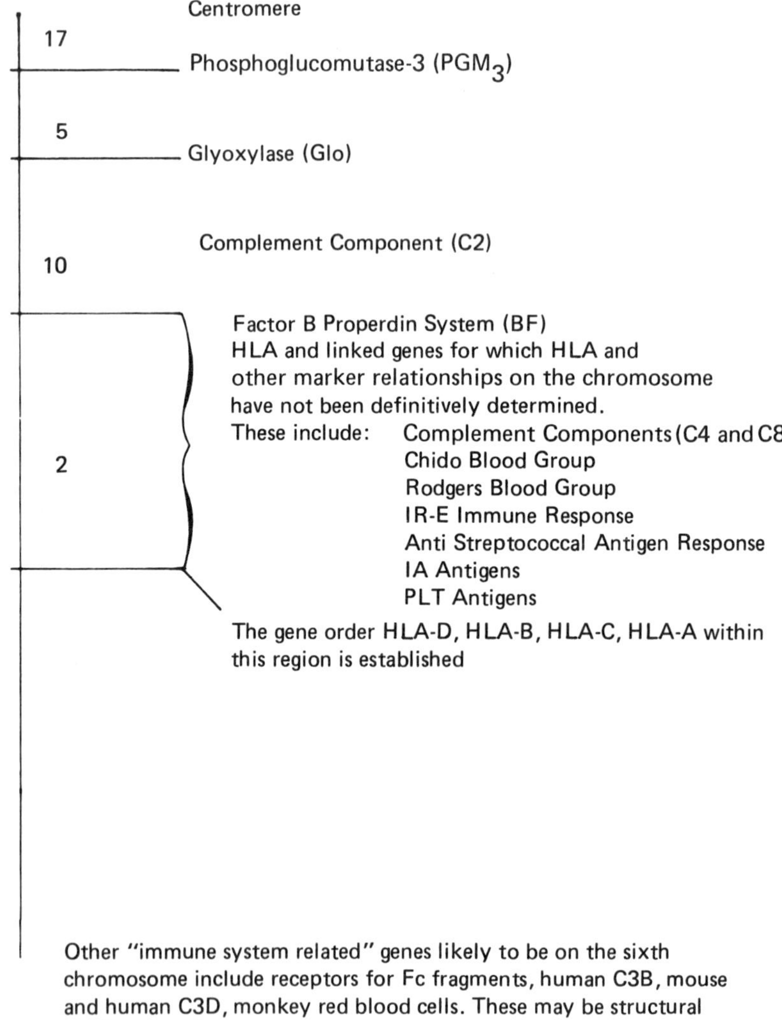

Figure 1. HLA and some other genetic relationships of the sixth human chromosome.

altered-self molecules, which might be involved in autoimmunity and tumor immunity. Thus, the extensive studies of immune response genes in rodents demonstrated that susceptibility to infectious agents or to autoimmune diseases might logically be expected to be associated with the HLA region in man on the basis of some individuals having less optimal constellations of immunoregulatory genes. As the association of HLA and numerous diseases became known, this predictable extension of experimental rodent immunogenetics to human biology became a reality. Nevertheless, despite extensive studies of the mechanism of immune response gene action in rodents, we still have no clear understanding of the mechanisms of HLA-associated human diseases.

Through the analysis of recombinant individuals, the MHC of both the mouse and man was soon divided into subgroups. The genes that code for the histocompatibility antigens themselves comprise at least two distinct loci. The region thought to contain the immune response genes and the most easily detectable mixed lymphocyte culture stimulating loci is distinct from the classical serological antigen determining genes. This major mixed lymphocyte reaction determining region is the *HLA-D* region in man and the *I* region of the mouse. The closest defined histocompatibility gene to *HLA-D* is *HLA-B,* and it is of significance that most of the HLA-associated diseases are more strongly associated with an HLA-B antigen. The other two defined HLA antigen genes—*HLA-A* and *HLA-C*—may also have associated diseases, but these are less definitively substantiated. In addition to the A, B, and C antigens, which are present on virtually all cells, B lymphocytes contain a separate series of antigens known as B cell antigen specificities which are genetically associated with the *HLA-D* locus. Since the *HLA-D* locus is thought to play the most predominant role in the regulation of immune responsiveness, we can expect an avalanche of disease associations with the newly defined B cell antigen specificities. These specificities, often referred to as Ia antigens because they are associated with the *HLA-D* region, which is analogous to the *I* region in the mouse, have recently been named *HLA-D* related B cell allospecificities.

Some of the genes of the MHC of man do not show random recombination in population studies. That is, several genes in the MHC are not in Hardy-Weinberg equilibrium. Another way to describe this fact is that these genes are in linkage disequilibrium. With the assumption of random mating and excluding recent mutation, any two linked genes should become randomized in populations through genetic recombination. The frequency of individuals or gametes with both genes should then be equivalent to the product of the individual gene frequencies. When this is not observed, there is said to be linkage disequilibrium for the two genes involved. Thus, an individual who is *HLA-A3* is more likely to be *HLA-B7* because A3 and B7 are in linkage disequilibrium; there are more A3, B7 gametes than would be predicted by the frequency of A3 and B7 gametes considered separately. This phenomenon of linkage disequilibrium or gametic association among particular specificities has been used as a possible explanation for disease associations with particular specificities as well. As discussed below, this interpretation may be formally correct, but it is not particularly useful to explain mechanisms. The phenomenon of linkage disequilibrium has also been responsible for a level of confusion with regard to the particular antigen associations of several HLA-associated diseases. For example, it is quite clear that in populations the HLA complex has some relationship to the disease multiple sclerosis. It is also known that *HLA-A3, B7, Dw2,* and the B cell specificities associated with Dw2 are in strong linkage disequilibrium. For example, a randomly selected *A3*$^{+}$ individual is more likely to be *B7*$^{+}$ than a randomly selected individual who is not *A3*$^{+}$. A similar argument holds for *B7* and *Dw2,* and for *Dw2* and particular B cell antigen specificities. The strength of this statistical observation increases from HLA-A to C to B and so on as we approach the B cell antigen specificity. Thus, it is most likely that an individual with *Dw2* will also have B cell antigen specificity DRw2. When the first studies of multiple sclerosis (MS) and HLA were conducted, a statistically significant increase of *HLA-A3* was found in multiple sclerosis patients. Subsequent studies demonstrated an increased frequency of B7, and the conclusion was drawn that the increase in A3 was a reflection of A3-B7 disequilibrium and that B7 was actually the important antigen associated with multiple sclerosis. The next development along this line was the finding that Dw2 was even more strongly associated with multiple sclerosis than was B7, leading to the conclusion that the B7 association was confounded by the association of B7 and Dw2. Later studies of B cell antigens associated with B7 and Dw2 demonstrated that nearly all multiple sclerosis patients contained certain B cell antigens, and the logical conclusion was then that the "real" multiple sclerosis-associated gene was strongly associated with the B cell antigens and all other associations were likely to be the result of the closest association. The final data for multiple sclerosis have not been completely analyzed to determine precisely what role A3, B7, Dw2, and even the B cell antigens themselves play in relation to a putative MS susceptibility locus of the HLA complex. In fact, family studies in which more than one multiple sclerosis patient is present in a given family do not always demonstrate segregation of MS with a particular HLA haplotype. Thus, we cannot yet determine whether there is a single HLA-associated genetic mechanism for associated diseases such as MS, or even whether single HLA specificities or *HLA* genes are by themselves responsible for statistically valid disease associations.

There are two basic approaches to the analysis of HLA and disease. The first would be to determine the frequency of particular HLA antigens among diseased and matched-control populations and the second would be to study the segregation of particular HLA haplotypes among families with multiple cases of a particular disease. Both population studies and family studies have a role to play in the analysis of HLA in disease. It is possible that a particular disease will be associated with an HLA antigen in the absence of the ability to study families with multiple cases, and it is equally possible that a disease may be HLA dependent in the absence of any statistically valid population association of the disease and a particular HLA specificity. It is important in the study of HLA and disease to be aware of the particular specificity or HLA loci being examined for a particular disease. For example rheumatoid arthritis has recently been shown to have a *HLA-D* locus association in the absence of any *HLA-B* locus association (3). This situation would have resulted in negative data for HLA association with rheumatoid arthritis in studies done before the last couple of years when HLA-D typing was available. Similarly, even though there are no known associations of particular HLA-A, B, or C antigens and ragweed allergy or leprosy, recent studies have shown the segregation of both diseases with an HLA chromosome within families (4,5).

In the application of this information to numerous diseases, it has been clear for some time that simple categorization of particular HLA types and associated diseases would not give significantly useful diagnostic or prognostic information for most HLA-associated diseases. Nevertheless, developments in basic immunobiology have been focused on the role of the MHC in the understanding of the regulation of the immune response. This latter development, which heralded the last decade of searches for HLA and disease associations, offers the possibility that the in-depth understanding of the mechanism of MHC gene action in experimental systems may further our understanding of human disease.

This chapter is not concerned with an attempt to categorize the voluminous literature of HLA association with disease. An international meeting was held approximately one year ago during which all the information available about HLA association to various categories of disease was assembled (6). The present chapter is an attempt to highlight some of the major issues that may be involved in determining disease mechanisms. We consider the relation of HLA and malignancy, HLA and autoimmunity, HLA and infectious diseases, and the possible role of HLA and the genetic determination of survival in the general sense, that is, HLA and aging.

HLA AND MALIGNANCY

The first experimental evidence to implicate the MHC in disease came from experimental work in mice which suggested that certain *H-2* types were more susceptible to the development of leukemia (7). Early work in HLA was directed to malignant disease partly in response to this seminal work in the mouse by Lilly, Boyse, and Old (7). Hodgkin's disease was investigated by Amiel (8) and Kourilsky and Dausset and their collaborators (9). These investigators studied the HLA profiles of patients with acute lymphocytic leukemia. Subsequent work in both Hodgkin's disease and ALL has supported the original suggestion of possible HLA associations.

A separate but equally important experimental and conceptual approach is the study of the role of the MHC in the host resistance to a transplanted or existing neoplasm. Just as HLA may be unrelated to the etiology of a virus or otherwise induced autoimmune disease, yet can still have dramatic influences over the pathogenesis of autoimmunity, the HLA complex may be involved in many steps in malignancy. Williams and his collaborators investigated the possibility that the phenomenon of enhanced tumor resistance of F_1 hybrids transplanted with parental strain tumors was a manifestation of MHC-linked immune response genes or immune response regulatory phenomena (10). It was demonstrated conclusively that more than one MHC gene may interact to produce alterations in host resistance to a single transplanted neoplasm. They also showed that such genes were not exclusively present in the *I* region of the mouse H-2 complex and thus must include genes that are distinct from the presently identified *H-2* linked immune response genes.

These two types of experimental approaches to the role of the MHC in malignancy, i.e., the MHC in relation to oncogenesis and the MHC in relation to tumor resistance, raise two of the many important considerations in regard to a role for HLA in malignancy. Beyond the consideration of oncogenesis or tumor resistance, the issue of tumor classification must be considered. Thus, the designation of particular HLA types in leukemia would be a broad classification that might obscure real HLA associations among subgroups of leukemia. Separation into categories of granulocytic and lymphocytic would also be inadequate. As tumors become more precisely categorized as to their progenitor cell origin, we may be able to relate specific HLA types to stages of malignant transformation.

An important consideration in human studies is the ethnic background of the patients examined. Referral patients to a cancer center may be strikingly different compared to the normal control population in that region. An analogy to experimental systems would be the known observation that leukemia virus may be located in different chromosome sites among various inbred strains of mice.

The important distinction between oncogenesis and tumor resistance has clear expression in HLA association of human malignancy. For example, we must consider the possibility that observed HLA associations to particular malignancies may represent survival differences rather than tumor incidence differences. This would be a particular problem because of the necessary retrospective character of HLA and malignancy studies. The most striking examples to be noted are an excess of *HLA-2* in long term survivors of acute lymphocytic leukemia and the excess of *HLA-Aw19* and *HLA-B5* among short-term survivors of Hodgkin's disease (11).

The currently most attractive single effector mechanism in tumor immunity with HLA influence on its function is the level of cell-mediated immunity. Studies in mice have suggested both an influence of the MHC on levels of natural killer cell activity, as well as the possibility that measured levels of natural killer cell activity can influence subsequent in vivo behavior of transplanted tumors (13). Natural killer cell levels seem to be elevated in F_1 hybrids (14). The possibility that natural killer cell activity may be involved in host survival should be given strong consideration, particularly in the case of leukemia in man. Published reports raise the possibility that *HLA-B12* individuals have relatively high levels of natural killer cell activity when tested on human cell lines (15). It has also been suggested recently that *HLA-B12* individuals have a better response to chemotherapy when they have the diagnosis of acute myelocytic leukemia (16,17), and acute leukemics have markedly depressed levels of natural killer cell activity (18). Thus, it may be that this effector mechanism which is known to be active against leukemia-derived cell lines in mice and humans may be regulated, in part, by genes associated with the MHC complex, and the consequence of this regulation may be reflected in the survival of patients with AML.

THE MAJOR HISTOCOMPATIBILITY COMPLEX, LYMPHOCYTE SUBSETS, AND MALIGNANCY

With the recognition of the complexity of immune response regulation as controlled by the MHC, we can now consider intervention in the host response to growing tumor through the selective elimination of particular lymphocyte subpopulations that express antigens of the major histocompatibility complex. The most straightforward example would be the attempt to suppress tumor growth by eliminating suppressor T cells that are functioning to inhibit the host's response to tumor. This approach has been taken by Green and his collaborators (19) and by Williams and Frelinger (20). Since the *I-J* region of the mouse MHC may be selectively expressed on suppressor cells, it was logical to attempt to influence tumor growth by anti-*I-J* serum injections. When histocompatible $I\text{-}J^k$-bearing animals were injected with anti-$I\text{-}J^k$ serum on days 0, 1, and 3 after giving one million P815 (no $I\text{-}J^k$ gene) cells, tumor sizes were significantly different, as noted in Table 1. Results of similar magnitude were reported by Green et al. using an antiserum that was also capable of eliminating suppressor cells and suppressor factor in the GAT system.

The antiserum-treated animals also survived longer than saline or normal mouse serum-treated controls. Experiments in progress also demonstrate that antisera directed against the *D*-region of the H-2 complex of the $H\text{-}2^d$ haplotype can result in suppressed tumor growth of a magnitude equal to or possibly greater than that achieved by the injection of anti-*I-J* serum. Since genetic control of host-resistance to P815 as a histocompatible tumor includes genes of the *D*-end of the H-2 complex, this latter approach may operate on the same cell subpopulations that are influenced by genes already known to play a role in host resistance. Further detailed analysis of this sort

Table 1. Effect of anti-I-J^k serum on P815 ($H\text{-}2^d$) tumor growth in $C3D2F_1$ ($H\text{-}2^{k/d}$) mice

	Tumor diameter (mm)			
	Day after injection			
Treatment of recipient	3	8	11	14
anti-$I\text{-}J^k$	0.08	0.73	1.25	1.75
Saline	0.18	1.09	1.75	2.65
p value	0.080	0.027	0.0007	0.0002

of experimental model will be necessary before any more specific conclusions may be drawn. However, these encouraging preliminary results raise the possibility that the production of antisera specific for human lymphocyte subpopulations may eventually have a therapeutic role in human malignancy.

HLA AND INFECTIOUS DISEASES

Although the evolution of interacting *HLA* genes probably was influenced by exposure of selected populations to infectious agents, there is very little current information relating the HLA system to specific infectious diseases. Most considerations of infectious disease and the MHC have been concerned with the possible influence of the MHC in the host response to infected cells rather than the response to the organism per se. The data of Zinkernagel and Dougherty and the extension of their observation that cytotoxic lymphocytes interact with virus antigens in association with or in the context of *H-2K* and *H-2D* gene products justifies the consideration of a possible role of HLA antigens themselves in HLA-associated diseases. Thus there may be specific HLA antigens that interact in unique ways with infectious agents to promote elimination, to generate indolent autoimmune disease through inadequate elimination, or to prevent elimination through inadequate host response. This latter possibility comprises the other most popular hypothesis for the role of HLA in infectious diseases, namely that of molecular mimicry. Particular HLA antigens may be identical to important specificities on infectious agents and thereby allow them to gain entry and to establish infection with relative permissiveness as the predominant host response. We consider this possibility to be very strong, yet it is unlikely that parasites, bacteria, or even viruses have so few antigenic specificities as to allow unimpeded infection and the complete absence of immune response. Rather, it is likely that molecular mimicry may enable relative increased infectivity of particular agents in specific HLA types with the manifestation of infection being dependent on the crossreactivity *and* nature of the host immune response. For example infection with the group A streptococcus can be associated with rheumatic fever and/or acute glomerulonephritis with the mechanisms apparently involving auto-aggressive immunity. Rheumatics have antistreptococcal antibodies that react with sarcolemma of human heart and human muscle cells. Mothers of rheumatics have a significantly lower incidence of *HLA-B5* (21). This is interesting because of the association between poor blastogenic response to purified streptococcal antigens and the absence of *HLA-B5* (*22*). Poor response to the streptococcus could result in repeated and prolonged infections and the subsequent development of nephritis or myocarditis. This is the case even in the absence of HLA associations in well-controlled studies of families of nephritics with rheumatic fathers. This possible maternal influence, in addition to the postulated HLA-associated genetic influence, serves to underscore the usual situation in the analysis of HLA and disease association, namely, the creation of suggestive mechanistic explanations and simultaneous unresolved exceptions.

HLA AND AUTOIMMUNITY

Three broad categories of HLA-associated diseases are heavily weighted with examples having suspected or proven immunological components in their pathogenesis. The *B27*-associated diseases, the most notable of which is ankylosing spondylitis, represent a category of disease in which spondolytic manifestations occur either in the defined syndrome of ankylosing spondylitis or in associated diseases such as Reiter's syndrome as a complication of inflammatory bowel disease, and even spondylitis as a consequence of infections with *Salmonella,* or *Yersinia.* In this case there is no known *HLA-D* allele association with *B27* and, likewise, there are no known *HLA-D* associations with the *B27*-associated diseases such as ankylosing spondylitis. The fact that approximately 90% of patients are positive for the *B27* allele compared to its frequency of only 5% in the general population suggests strongly that the B27 antigen itself may play some role in the disease process.

Diseases such as multiple sclerosis (MS) and juvenile diabetes mellitus (JDM) have a relatively small numerical predominance of specific *D* locus specificities, yet the statistical significance of association is beyond doubt. In the case of JDM both *B8* and *BW15* have an increased incidence and the *DW3* allele is even more frequently associated with the disease. Of interest is the fact that *B8* or *BW15* homozygotes are not at increased risk as compared to individuals with only one *B8* or *BW15* allele. Nevertheless, the combination of *B8* and *BW15* does produce an additive risk. This situation suggests that more than one MHC gene in the *trans* configuration can produce increased disease susceptibility, even though having twc doses of a single allele in the homozygous form does not result in increased risk. Similar studies of the relative influence of the different known associated HLA specificities in MS and JDM are presently being conducted. A major issue in this sort of disease is whether or not there are additive influences of multiple MHC genes. In addition, there is an interesting reciprocal relationship between *B* and *D* locus genes in MS and JDM. The *B7* and *DW2* genes, which are relatively increased in multiple sclerosis, are relatively absent in diabetes. This is similar to the relative decrease in *HLA-B12* among MS patients. As discussed below, such results raise the possibility that MHC genes may interact in both a positive and negative sense in the expression of a disease.

Individuals with gluten-sensitive enteropathy or dermatitis herpetiformis or hypocomplemententemic glomerulonephritis seem to possess a receptor that is expressed on B lymphocytes. This receptor may or may not be under genetic control, but its expression may be related to the *HLA* allele of the individual. Thus, in gluten-sensitive enteropathy, it may be that all infected individuals react with a gliadin receptor and that the great majority, but not all, have the *DW3* allele. Therefore it would seem that two B cell alloantigens may be involved in gluten-sensitive enteropathy; one associated with *HLA-DW3* and one probably not *HLA* related.

Consideration must always be given to the frequency of *HLA* alleles in disease association among different populations. Myasthenia gravis is associated with *B12* and *B35* in Japanese, whereas it is associated with *B8* in Caucasians. Conversely, ankylosing spondylitis is associated with *B27* in the Japanese and Caucasian population, and *B5* is associated with Bachet's disease in Japanese, Europeans, and Mideastern people. A third example is the association of *BW22.2* with thromboangiitis obliterans and arteriosclerosis obliterans in Japanese, even though the diseases have no Caucasian *HLA* association.

LINKAGE DISEQUILIBRIUM AND MULTIPLE INTERACTING MHC GENES IN DISEASE ASSOCIATIONS

Linkage disequilibrium in the HLA system was first noticed when certain *HLA-A* specificities were found to co-exist on the haplotype of a given *HLA-B* specificity in populations, for example, *A1* with *B8*, *A3* with *B7*, or *A2* with *B12*. Such disequilibrium includes *HLA-D* and may be extremely high between *B* and *D*, for example, *DW2* and *B7* or *DW3* and *B8*. The possibility that present-day Caucasians represent descendants from diffusion of a number of small and previously inbreeding populations has been raised to explain observed linkage disequilibrium. Noted differences between Caucasoid and Japanese HLA antigen frequencies may have resulted from gene drift and neutral or natural selection. The contribution of geographic isolation in earlier centuries and inbreeding can produce differences in haplotype structure. The often quite distinctive Caucasoid and Oriental haplotypes may be explained by this geographic isolation. The differences will involve the antigens we identify, *HLA-A, B, C,* and *D,* and also loci that are present on the haplotype and may be involved in the predisposition to disease for which there are no simple means of identification. These genes may include immune response genes, immunosuppression genes, and genes that code for virus or antigen receptors, or even genes that code for DNA sequences that are sites of virus integration. A more critical analysis of the role of *HLA* in association with disease would include the possibility that multiple *HLA*-linked genes play a role in a single disease. The phenomenon of linkage disequilibrium or gametic association may be a result of the evolutionary advantage for maintaining multiple interacting immune response genes within the MHC. Although the evolutionary pressures that resulted in linkage disequilibrium are unlikely to be discernable among today's human population, there may still be remnants of this phenomenon through the immunological processes that depend on the interaction of multiple MHC components. The phenotype of a gene cluster that may have been advantageous for survival in the past may now be expressed in a way that we interpret as a disease. Infectious diseases may have acted not only to eliminate individuals less equipped to fight them, but they may also have produced during evolution an arrangement of interactive genes that would have survival value. Such gene clusters may have been important for resistance to an ancient bacteria or virus, but later mutations in the infectious agent or the host may now be interpreted as unexplained linkage disequilibrium and disease.

An alternative explanation for linkage disequilibrium is that certain gene combinations function optimally in the *cis* configuration. Thus, the selective advantage that we postulate for them may have been present only when the individual genes were inherited on the same chromosome. Since this consideration need only apply to selected loci, the combination of inbreeding and outbreeding in the context of natural selection could work together to produce specific gene clusters. A separate explanation for linkage disequilibrium is that a particular characteristic such as an allele of the *T*-locus in the mouse may be present on the same chromosome as the MHC. This linked gene may increase the lifespan of sperm in favor of the transmission of particular MHC clusters even when they are deleterious. Another explanation is that there is no mechanistic explanation and one gene "hitch-hikes" with the other over the time period available to examine.

The comparison of haplotypes between different ethnic groups might help to resolve the nature of linkage disequilibrium. The *HLA-D* specificity *LDHO* which is in disequilibrium with *A8, BW35* is absent in Caucasians but is found associated with myasthenia gravis and Grave's disease among Japanese. *HLA-DYT* is found in linkage disequilibrium with *A9, BW22* and is absent from Caucasians but is found in association with juvenile onset diabetes mellitus in Japanese. Thus, if mutations occurred in ancestral man to explain susceptibility to myasthenia gravis and Grave's disease or diabetes mellitus, they should have occurred in a portion of chromosome 6 close to *DW3* in Caucasians but close to *HLA-DLDHO* and *HLA-DYT* in Japanese. In this same context, the most common haplotype in association with ankylosing spondylitis in Orientals and Caucasians is *A2, B27. A1, B27* and *A3, B27* are Caucasoid haplotypes, but *A2, B27,* is not. This suggests that the ankylosing spondylitis carrying *B27* haplotype, *A2, B27,* is an example of the segregation of an Oriental haplotype in a Caucasian population.

The phenomenon of incomplete penetrance in HLA-associated disease is rather easily accommodated by our hypothesis that multiple *MHC* genes may be required to produce specific disease syndromes. The combination of several genes in the *MHC* may have been required to eliminate an infectious agent during human evolution, whereas in more modern times the same gene cluster may promote the development of autoimmune disease. The autoimmune manifestation may depend on reactivity to altered self (? HLA antigen), while the particular infectious agent threat originally responsible for the maintenance of a particular gene cluster may no longer be present in its original form. The random separation of interacting genes in a particular autoimmunity producing gene cluster would leave the individual carrying these genes apparently unaffected, whereas the offspring of such individuals may inherit complementing genes that would allow expression of the disease. Disease expression in this instance would require the ability for genes to interact in the *trans* configuration, and both parents would contribute in the genetic sense to the child's susceptibility. Likewise, a single individual who carries components of required interacting gene clusters inherited from both his parents may not manifest severe disease if the interacting gene clusters required *cis* configuration for full expression. The appropriate recombinant event in this individual would thus generate a single haplotype with complete ability to manifest the disease, yet it would only be seen in a subsequent generation.

IMMUNOGENETICS OF HLA AND AGING

Whether life is terminated by degenerative disease or by the imposition of fatal infectious disease on the background of chronic disease, longevity per se may be related to the balance of immune reactions to the environment, both external and internal, and the genetic regulation of the immune responses that are necessary for disease resistance. Studies of immunogenetics and aging are just being undertaken in man. Of great interest and potential for the study of human immunogenetics and aging is the observation that *HLA-B8* is decreased in older females and that the remaining *B8* females have lower *PHA* responses than matched *B8* negative controls (23). This raises the possibility that decreased longevity may be accompanied by a deficiency in T cell function which may be under detectable genetic control.

MECHANISMS OF HLA AND DISEASE ASSOCIATION

Theories of the importance of HLA antigens or associated MHC genes in disease fall into two categories. In the first the putative disease gene is located in the HLA complex, but is distinct from any other defined *HLA* genes. Alternatively, the disease susceptibility gene may be unlinked to *HLA*, but the disease itself may be modified by MHC genes. Genetic diseases are frequently multifactorial and HLA may be one of many genetic factors involved.

If the disease susceptibility gene is present in the MHC, then it may logically be there simply by chance or it may function in the context of its associated MHC genes. The chance association hypothesis cannot be formally tested and represents a restatement of the fact of linkage disequilibrium among *HLA* loci.

We believe that at least two mechanisms and possibly many more are involved in autoimmune susceptibility. The first would involve participation of HLA antigens or other receptors coded for by the HLA region. These may be single factors expressed co-dominantly or multi-gene complexes that may function in *cis* configuration when inherited from a single parent or by complementary *trans* interactions with participation of both parental haplotypes, either of which may be modified by intrinsic (metabolic) errors or extrinsic (viruses, chemicals, etc.) factors. Abnormal immunoregulation involving cellular or molecular events controlled by interacting MHC genes is the other general mechanism likely to be involved in MHC-associated diseases.

Experimental animal systems will be required to elucidate more precisely the genetically controlled mechanisms involved in the MHC association with disease. We have already mentioned the use of F_1 hybrids with defined *MHC* genotypes in the analysis of the role of HLA and malignancy (24). In autoimmune disease, experimental allergic encephalomyelitis (EAE) in the inbred rat is the best defined model and is an excellent system in which to analyze the influence of the MHC on disease susceptibility and severity. EAE is an autoimmune disease that can be elicited by the injection of a defined polypeptide known to be present in the central nervous system myelin. Susceptibility to the clinical or pathological expression of this autoimmune disease in the rat depends on the presence of a single *MHC*-linked gene. Lewis rats are completely susceptible; BN rats are uniformly resistant; and the (LEW × BN) F_1 is entirely susceptible, but develops a disease of diminished severity. The susceptibility gene has been termed *Ir-EAE* and has been shown to be linked to the MHC of the rat (25). Recent studies have demonstrated that the severity of the EAE syndrome, independent of susceptibility, is also dependent on the MHC. In segregating F_2 and backcross populations, animals that are homozygous for the MHC of the Lewis type have more severe disease than animals that are heterozygous. The nonsusceptible BN rat may contribute immunoregulatory genes that are expressed as the suppression of the severity of the disease. An alternative explanation is that the nonsusceptible BN-like MHC, which by definition is *Ir-EAE* negative, may actually contribute an immunosuppression gene termed *Is-EAE* which prevents full development of the disease. The possibility remains that BN rats are phenotypically nonsusceptible as a consequence of the overactivity of or abundance of suppressor genes and that the simple lack of "responder genes" is an inadequate explanation for the nonsusceptible individual. To distinguish between the alternatives of gene dose effect of two *Ir-EAE* genes compared to the disease suppressing effect of a putative *Is-EAE* gene will require the generation of MHC recombinants between susceptible and resistant haplotypes. It will then be possible to determine whether partial *MHC* heterozygotes can ever develop as severe an EAE syndrome as the Lewis-type *MHC* homozygote. This analogy to disease susceptibility in combination with disease severity may be made in man in multiple sclerosis. *HLA-B12* is selectively absent from MS patients. Thus it is conceivable that an *HLA-B12*-associated immunoregulatory gene (*?Is-MS*) may exist. The function of this hypothetical gene would be to prevent the manifestation of MS symptoms even in the presence of an otherwise genetically susceptible and environmentally exposed individual (i.e., *Ir-MS*$^+$). It would be informative to determine whether the presence of *HLA-B12* in association with *HLA-DW2* and/or *HLA-B7* leads to less severe multiple sclerosis compared to *B12* negative, *DW2,* or *B7* positive individuals. It should be emphasized that the MHC influence may be restricted to the pathogenesis of disease and may have absolutely nothing to do with its etiology. For example, multiple sclerosis might well be caused by virus, but the lesions may result from an inappropriately regulated immune response to the virus, the myelin, the histocompatibility antigens, or some combination of all three.

Experimental studies of immunogenetics and aging may also be pursued fruitfully in rodent systems. Various inbred strains of mice have dramatically different longevity. For example, the NZB and (NZB $\times$ NZW)F_1 mice have markedly shortened survival, which could be based on immunologic function. The fact that some NZB hybrids such as (NZB $\times$ C3H)F_1 mice live longer than the NZB parental strain suggests that heterozygosis may on the one hand increase susceptibility to disease and early demise, whereas heterozygosis with the right gene combination may result in prolonged survival. The human analogy in this case is that certain studies have shown *HLA-B* locus heterozygotes to be more frequent in the elderly than in young cohorts. We have already noted that *B8* is decreased among older females. However, *B8* is also associated with increased frequency of autoimmune disease. The *HLA-A1, B8* haplotype which is associated with decreased survival under most circumstances is also associated with increased survival when the individuals concerned have cancer of the breast. The paradox is likely to be explainable by the postulate that autoimmunity may generate tumor immunity. Since the *HLA-A1, B8* haplotype is associated with a generally increased tendency toward immunity, it may also have enhanced tumor immunity and may aid in the survival of individuals with breast cancer. It should be noted that this postulated increase in survival would have nothing to do, necessarily, with tumor susceptibility.

Although tumor development is often considered to be a phenomenon of the old and the issue of the role of decreasing immune function has been raised, there are genetic circumstances under which tumors grow less well in aged individuals as compared to genetically identical younger animals. For example, aged $B6D2F_1$ mice have more slowly growing transplanted P815 tumors as compared to aged $C3D2F_1$ mice. This finding may depend on multiple mechanisms having to do with tumor immunity, immunogenetics, and aging, but it does serve to illustrate the use of experimental models to approach such questions.

Studies of HLA and disease associations shed considerable light on our understanding of both immunogenetics and human disease. It is likely that multiple interacting MHC genes influence host response to environmental insults as well as internal deregulation of homeostatic mechanisms. Precise understanding of specific disease categories will almost certainly require development of model systems in experimental animals with the subsequent application of principles and hypotheses derived to the data possible to collect among human populations. The epidemiological human data may ultimately allow identification of at-risk populations even in situations where preventive measures may be applicable. However, the use of HLA and disease association knowledge will probably depend on information first developed in experimental systems. It is encouraging that the development of basic immunobiology in relation to the MHC has kept pace with the past decades' documentation of previously predicted associations of the MHC with disease. The next decade of HLA and disease studies should focus on studies that may point to mechanistic interpretations of the pathogenesis of selected HLA-associated diseases as we continue to extend our knowledge of the major histocompatibility complex.

LITERATURE CITED

1. References to original or confirming references relating specific HLA genes to disease are not included in this review, which is concerned with general concepts relating to possible mechanisms. Several detailed reviews of HLA, including disease associations, with numerous references include:
   Amos, D. B., and Pool, P. 1976. HLA Typing. In N. R. Rose and H. Friedman (eds.), pp. 797–804. Manual of Clinical Immunology, American Society for Microbiology, Washington, D.C.
   Bach, F. H., and van Rood, J. J. The major histocompatibility complex—genetics and biology, New England J. Med. 295:806–813.
   Dausset, J., and Svejgaard, A. (eds.). HLA and Disease, p. 333. Institut National de la Sange Ed de la Recerdue Medicale, Paris.
   Dupont, B., Hansen, J. A., and Yunis, E. J. 1976. Human mixed lym-

phocyte culture reaction: Genetics, specificity and biological implications. Adv. Immunol. 23:107-122. Academic Press, New York.
Lange, C. F. 1975. HL-A histocompatibility antigens and their relation to disease. Prog. Clin. Pathol. 137-158.
Moller, G. (ed.). HLA and disease associations—A survey. Transpl. Rev. 22:3-195.
Ryder, L. P., and Svejgaard, A. (eds.). 1976. Report from the HLA and Disease Registry of Copenhagen. 34 pages.
Sasazuki, T., Grumet, F. C., and McDevitt, H. O. The association between genes in the major histocompatibility complex and disease susceptibility. Ann. Rev. Med. In press.
Svejgaard, A., Hauge, M., Jersild, C., Platz, P., Ryder, L. P., Staub-Nielsen, L., and Thomsen, M. 1975. The HLA system: An introductory survey. Monogr. Hum. Genet. 7:3-102.

2. Cannady, W. G., DeWolf, W. C., Williams, R. M., and Yunis, E. J. 1977. Laboratory methods in transplantation immunity. Adv. Clin. Pathol. 7:239-265.
3. Stasny, P. 1974. Mixed lymphocyte typing cells from patients with rheumatoid arthritis. Tissue Antigens 4:571.
4. Blumenthal, M. N., Amos, D. B., Noreen, H., Mendell, N. R., and Yunis, E. J. 1974. Genetic mapping of Ir gene in man. Linkage to second locus of HLA. Science 184:1301-1303.
5. de Vries, R. R. P., Fat, L. A., Nigenhuis, L. E., and van Rood, J. J. 1976. HLA-linked genetic control of host response to mycobacterium leprae. Lancet ii (7999):1328-1330.
6. Dausset, J., and Svejgaard, A. (eds.). 1977. HLA and Disease. Munksgaard, Copenhagen. Williams & Wilkens, Baltimore.
7. Lilly, F., Boyse, E. A., and Old, L. J. 1964. Genetic basis of susceptibility to viral leukemiogenesis. Lancet ii:1207-1209.
8. Amiel, J. L. 1967. Study of the leukocyte phenotypes in Hodgkin's disease. In E. S. Curtoni, P. L. Nattiuz, and R. M. Tosi, (eds.), Histocompatibility Testing, pp. 79-81. Munksgaard, Copenhagen.
9. Kourilsky, F. M., Dausset, J., Feingold, N., Dupuy, J. M., and Bernard, J. 1967. Leukocyte groups and acute leukemia. J. Nat. Cancer Inst. 41: 87.
10. Williams, R. M., Dorf, M. E., and Benacerraf, B. 1975. H-2 linked genetic control of resistance to histocompatible tumor. Cancer Res. 35: 1586-1590.
11. Falk, J., and Osoba, D. 1977. The HLA system and survival in malignant disease: Hodgkin's disease and carcinoma of the breast. In G. P. Murphy (ed.), HLA and Malignancy, pp. 205-216. Alan R. Liss, New York.
12. Petranyi, G. G., Kiessling, R., Povey, S., Klein, G., Herzenberg, L., and Wigzell, H. 1976. The genetic control of natural killer cell activity and its association with *in vivo* resistance against a moloney lymphoma isograft. Immunogenetics 3:15-28.
13. Williams, R. M., Leifer, J., and Moore, M. J. 1977. Hybrid effect in natural cell mediated cytotoxicity to SV40 transformed fibroblasts by rat spleen cells. Transplantation 23:283-286.
14. Trinchieri, G., Santoli, D., Zmijewski, C. M., and Koprowski, H. 1977.

Functional correlation between antibody-dependent and spontaneous cytotoxic activity of human lymphocytes and possibility of an HLA-related control. Transpl. Proc. 9:881-884.
15. Oliver, R. T. D., Klouda, P., and Lawler, S. HLA associated resistance factors and myelogeneous leukemia. In press.
16. Parrish, E. J., Heise, E. R., and Cooper, M. R. 1977. HLA association with acute myelogeneous leukemia. In G. P. Murphy (ed.), HLA and Malignancy, pp. 81-89. Alan R. Liss, New York.
17. Pross, H. F., and Baines. M. G. 1976. Spontaneous human lymphocyte-mediated cytotoxicity against tumour target cells. I. The effect of malignant disease. Int. J. Cancer 18:593-604.
18. Green, M. I., Dorf, M. E., Pierres, M., and Benacerraf, B. Reduction of syngeneic tumor growth by an anti-I-J alloantiserum specific for suppressor T cells. Proc. Nat. Acad. Sci. In press.
19. Williams, R. M., and Frelinger, J. A. 1978. Regulation of histocompatible tumor growth by antiserum prepared against subregions of the MHC. Cancer Res. 19:104.
20. Read, S. E., Reid, H., Poon-King, T., Fischetti, V. A., Zabriskie, J. B., and Rapaport, F. T. 1977. HLA and predisposition to the nonsuppurative sequelae of group A streptococcal infections. Transpl. Proc. 9:543-546.
21. Greenberg, J., Gray, D., and Yunis, E. J. 1975. Association of HLA-A5 and immune responsiveness *in vitro* to streptococcal antigens. J. Exp. Med. 141:935-943.
22. Hallgren, H. M., Kersey, J. H., Dubey, D. P., and Yunis, E. J. 1978. Lymphocyte subsets and integrated immune function in aging humans. Clin. Immunol. Immunopathol. 10:65-78.
23. Williams, R. M. 1977. Experimental models with possible implications for the role of HLA in malignancy. In G. P. Murphy (ed.), HLA and Malignancy, pp. 21-28. Alan R. Liss, New York.
24. Williams, R. M., and Moore, M. J. 1973. Linkage of susceptibility to experimental allergic encephalomyelitis to the major histocompatibility locus in the rat. J. Exp. Med. 138:775-783.

Infection, Immunity, and Genetics
Edited by Herman Friedman, T. Juhani Linna, and James E. Prier

MULTIPLE SCLEROSIS, THE HLA SYSTEM, AND VIRUSES

Donald H. Silberberg

Multiple sclerosis (MS) is the most common chronic disease of the central nervous system among adults in North America and Europe. MS affects approximately 50 per 100,000 population in these areas. It has been recognized as an entity for over 100 years, but despite our clinical and neuropathologic familiarity with MS, we are just beginning to achieve some understanding of its pathogenesis. It seems probable that its cause will prove to be multifactorial and involve immune alterations and/or viral infection of genetically susceptible individuals. The end result of the disease, which leads to its name, is the production of multiple areas of demyelination within the brain and spinal cord. Astrocytic proliferation within demyelinated plaques leads to firm scarring or sclerosis. These lesions are associated with the development of dysfunction of those body parts supplied by the particular area of the central nervous system in which demyelination is produced.

Epidemiologic studies of the prevalence of MS related to latitude were the first sets of observations that provided an important clue in understanding MS in those countries where MS has a relatively high prevalence rate. This seems to be limited primarily to Caucasian populations, where the prevalence rates are consistently higher further from the equator than closer to the equator. For example, MS is three to four times more common in Boston or Winnepeg than in New Orleans. This distribution provided the basis for carrying out a number of prevalence studies among migrants to low risk areas. Although the actual numbers are small, the rate among immigrants to Israel and to South Africa from high and low risk areas provided evidence that an immigrant acquires the low risk of Israel or South Africa (1/100,000) if migration occurs in the first several years of life. If immigration occurs around puberty or later, the migrating population carries with it the same risk as the country of origin. Closer study of these data suggested an approximately 15-year latency between exposure to an environmental factor and the first clinical evidence of MS (1).

More recent support for the importance of environmental exposure is the apparent cluster of MS cases that occurred in the Faroe Islands in the 1940s through early 1960s. Appearance of MS followed occupation of the Islands by British Marines during World War II (2). Further studies are needed to clarify the factors involved in this significant experiment of nature.

Population studies can also be used to make a strong case for involvement of a genetic factor in the development of MS. The islands of Japan occupy north temperate latitudes that are similar to those of the United States from Boston to Charleston, South Carolina. Modern health facilities are available to the majority of the Japanese population. Despite this, the prevalence of MS in Japan is very low, from 1–3 per 100,000. Recent studies carried out in both the Seattle and Los Angeles metropolitan areas show that first and second generation Japanese-Americans retain their low incidence of MS, suggesting a genetically determined lack of susceptibility to MS (3).

HLA studies are beginning to provide a rationale for understanding these epidemiologic studies. Early studies showed a higher frequency of *HLA-A3* and *HLA-B7* among MS patients, as compared with suitable controls, in Caucasian populations (4,5,6). An even more impressive association was found between the locus *HLA-Dw2*, with approximately 60% of MS patients expressing *Dw2* as compared with 15% of controls (7,8). More recently certain B lymphocyte determinants or alloantigens have been found in approximately 80–90% of patients with MS, as compared with 10–35% of the normal population, in studies carried out in the U.S. (9,10), England, and Jordan (11). The study comparing patients in England with clinically indistinguishable patients in Jordan showed a strong association between two different B lymphocyte determinants in the two populations (11). A possibility is that both alloantigens, and perhaps those reported in other studies, are in linkage disequilibrium with a single, as yet unidentified, susceptibility gene. Another possibility is that each separate gene confers susceptibility to MS in the appropriate population, but that the environmental factor in different regions varies.

Additional evidence for the importance of genetic susceptibility derives from a study of Black American patients with multiple sclerosis. Previous studies showed that the frequency of *HLA-A3* and *HLA-B7* in East African Black populations is very low. We reasoned that this indicated the possibility of a low frequency of *HLA-Dw2*. A comparison of 31 MS patients in Philadelphia with 34 controls showed that 11 of the 31 patients expressed *Dw2*, whereas none of the 34 controls did (12).

Before studies of the HLA system, it had become clear that the frequency of MS among immediate relatives of patients with MS was five to ten times higher than what would be expected by chance alone. Studies of identical twins showed afflication of the second twin in 18% of instances (1).

Evidence for involvement of a virus or other infectious agent in the pathogenesis of MS remains indirect. The time-tested finding of elevated levels of cerebrospinal fluid IgG in patients with MS (13), and the more recent discovery that most patients show an oligoclonal pattern when this gamma region is electrophoresed on agar or agarose (14), associates MS with other diseases with similar findings. These include diseases of known infectious etiology, such as meningitis, encephalomyelitis, tertiary syphillis, and at least one disease thought to be the result of persistent virus infection, subacute sclerosing panencephalitis (SSPE) (15). Most of the cerebrospinal fluid IgG seems to be synthesized within the central nervous system (16). A small part represents antibodies to measles, and perhaps to other viral antigens (17). The target antigen of the bulk of the CSF or IgG remains to be determined.

Since it was first described in 1962, the existence of a slightly raised titer of measles antibody in groups of MS patients, as compared with normals, has been confirmed and extended to similar observations for vaccinia, herpes, and paramyxoviruses. The possible relationship between antibodies to measles or other viruses and HLA haplotype is under active investigation. Several conflicting studies have been published that leave open the question of whether or not such a relationship does exist (18). The recent serologic studies of the populations of the Orkney and Shetland Islands, where the prevalence of MS is the highest in the world, show no differences in viral antibody titers between patients and controls. This mitigates against involvement of any of these agents as the significant or sole cause of MS (19).

A paramyxovirus was isolated from the brains of two patients with MS, but in a relatively small number of attempts to confirm this observation the finding has not been reproduced (20). Nucleocapsid-like tubules have been seen in macrophages in the vicinity of MS lesions and in the brains of patients with other CNS diseases (21,22). Their significance is unclear.

In summary, epidemiologic data support the operation of both an environmental agent and genetic susceptibility in the pathogenesis of MS. The HLA system is beginning to provide a rational basis for understanding the genetic susceptibility, and may serve to explain the mechanism by which particular individuals develop the recurrent demyelination that characterizes MS.

LITERATURE CITED

1. Leibowitz, U., and Alter, M. 1973. Multiple Sclerosis: Clues to Its Cause. North-Holland, Amsterdam.
2. Kurtzke, J., and Hyllested, K. 1975. Multiple sclerosis: An epidemic disease in the Faroes. Trans. Am. Neurol. Assoc. 100:213-215.
3. Detels, R., Visscher, B., Coulson, A., and Malmgren, R. 1975. Neurology 25:357.
4. Naito, S., Namerow, N., Mickey, M. E., and Terasaki, P. I. 1972. Multiple sclerosis: Association with HLA-3. Tissue antigens. 2:1-4.
5. Bertrams, J., Kuwert, E., and Liedtk, E. 1972. HLA antigens in multiple sclerosis. Tissue Antigens 2:405-408.
6. Jersild, C., Svejgaard, A., Fog, T., and Amitzboll, T. 1973. HLA antigens and diseases. II. Multiple sclerosis. Tissue Antigens 3:243.
7. Jersild, C., Fog, T., Hansen, G. S., Thomson, M., Svejgaard, A., and Dupont, B. 1973. Histocompatibility determinants in multiple sclerosis, with special reference to clinical course. Lancet 2:1221-1225.
8. Opelz, G., Terasaki, P., Myers, L., et al. 1977. The association of HLA antigens A3, B7, Dw2 with 330 multiple sclerosis patients in the United States. Tissue Antigens 9:54-58.
9. Terasaki, P., Park, M. S., Opelz, G., et al. 1976. Multiple sclerosis and high incidence of a B lymphocyte antigen. Science 193:1245-1247.
10. Winchester, R. J., Ebers, G., Fu, S. M., et al. 1975. B-cell alloantigen Ag 7a in multiple sclerosis. Lancet 2:1221.
11. Kurdi, A., Ayesh, I., Abdallat, A., Maayta, U., McDonald, W. I., Compston, D. A. S., and Batchelor, J. R. 1977. Different B lymphocyte alloantigens associated with multiple sclerosis in Arabs and N. Europeans. Lancet 1:1119-1121.
12. Dupont, B., Lisak, R. P., Jersild, C., Hansen, J. A., Silberberg, D. H., Whitsett, C., Zweiman, B., and Ciongoli, K. 1977. HLA antigens in black American patients with multiple sclerosis. Transplant. Proc. 9 (suppl. 1):181-185.
13. Kabat, E. A., Glasman, M., and Knaub, E. 1948. Quantitative estimation of the albumin and gamma globulin in normal and pathological cerebrospinal fluid by immunochemical methods. Am. J. Med. 4:653-662.
14. Laterre, E. C., Callewaert, A., Heremanns, J. F., et al. 1970. Electrophoretic morphology of gammaglobulins in cerebrospinal fluid of multiple sclerosis and other diseases of the nervous system. Neurology 20:982-990.
15. Link, H., and Muller, R. 1971. Immunoglobulins in multiple sclerosis and infections of the nervous system. Arch. Neurol. 25:325-344.

16. Tourtellotte, W. W. 1971. Cerebrospinal fluid immunoglobulins and the central nervous system as an immunological organ particularly in multiple sclerosis and subacute sclerosing pancephalitis. In L. P. Rowland (ed.), Immunological Disorders of the Nervous System. Williams & Wilkins, Baltimore. Res. Publ. Assoc. Res. Nerv. Ment. Dis. 49:112-147.
17. Norrby, E., Link, H., Olsson, J. E., et al. 1974. Comparison of antibodies against different viruses in cerebrospinal fluid and serum samples from patients with multiple sclerosis. Infect. Immun. 10:688-694.
18. McFarlin, D. E., and McFarland, H. F. 1976. Histocompatibility studies and multiple sclerosis. Arch. Neurol. 33:395-398.
19. Poskanzer, D. C., Sever, J. L., Terasaki, P. I., Prenney, L. B., and Sheridan, J. L. 1977. Multiple sclerosis in the Orkney and Shetland Islands. II. Viral antibody titers, history of infection, and etiology. Neurology 27:372.
20. Koprowski, H. 1976. Search for viruses in multiple sclerosis. Neurology 26:81-82.
21. Prineas, J. 1972. Paramyxovirus-like particles associated with acute demyelination in chronic relapsing multiple sclerosis. Science 178:760-763.
22. Tanaka, R., Iwasaki, Y., and Koprowski, H. 1976. Paramyxovirus-like structures in brains of multiple sclerosis cases. Arch. Neurol. 32:80-83.

Infection, Immunity, and Genetics
Edited by Herman Friedman, T. Juhani Linna, and James E. Prier

MURINE LUPUS ERYTHEMATOSUS AND ITS CLINICAL IMPLICATIONS

Ann E. Gabrielsen

The (NZB × NZW)F_1 hybrid mouse (B/W) develops antinuclear antibodies early in life, followed by a rapidly progressive glomerulonephritis (1) in which the dominant immune complex is DNA-anti-DNA (2). The disease bears a striking resemblance to that of many systemic lupus erythematosus (SLE) patients, and investigations of the mouse model have had a major impact on current concepts of SLE, particularly those implicating C-type RNA tumor viruses (3-5) and inadequate suppressor cell function (6).

There had been evidence, from major laboratories, that ecotropic viruses of the Friend-Moloney-Rauscher (FMR) and Gross groups, as well as a xenotropic virus, were present in the NZB and its hybrids, including the B/W (3-5). Several additional laboratories have been unable to confirm the presence of the ecotropic viruses, however (7,8), and the best sustained claim at present is the presence of the xenotropic virus in substantial titer in the NZB and its hybrids (7,9). It has not been possible to correlate the titer of the virus with autoimmune manifestations (9), and it is not clear that this virus has any central role in either NZB or B/W disease.

The issue of the status of suppressor cells in the B/W mouse is also unresolved, since there is evidence of both too little (6,10) and too much (11) suppression in these animals. Thus far, repeated thymus grafting or periodic administration of large inocula of dispersed lymphoid cells, both from young syngeneic mice with adequate suppressor populations, have only marginally increased survival of B/W mice (12,13).

Very little attention has been given to the problem of the source of the DNA in the immune complexcs in B/W renal glomeruli. If the mechanism of immune complex deposition in these animals is the classic one, antigen must be available in excess to the homologous antibodies over an extended period of time. A basic defect in the B/W mouse may be inappropriate availability of substantial volumes of nuclear materials from maturing erythrocytes or keratinizing skin cells (14). Successful treatment of these animals with drugs has usually been considered immunosuppressive or anti-inflammatory or both, but these agents could be inhibiting input of antigen since most of them affect all rapidly proliferating cell populations. The most effective drugs—prednisone, azathioprine, and cyclophosphamide (6)—all suppress erythropoiesis in mice to varying degrees (15).

Our therapeutic approach has been to suppress erythropoiesis as selectively as possible and to document the effects on progress of the renal disease. We chose a physiologic means initially, seeking to avoid the ambiguities inherent in the pharmacologic approach. Hypertransfusion has been widely used to suppress erythropoiesis in erythropoietin assays, and it also prolongs survivals of Friend and Rauscher leukemia virus-injected mice substantially (16,17). We prolonged survivals of B/W female mice significantly, but the mean increase in survival was limited to about 20% (18). We did not use the very intense transfusion regimen that was successful in the Friend and Rauscher leukemias, and it may be that such a schedule would be more beneficial. The transfusion experiments overlapped with early dactinomycin work and were discontinued when the efficacy of the drug became evident.

Dactinomycin (actinomycin D) is also used to suppress erythropoiesis selectively in vivo. When given in low doses, it will inhibit ^{59}Fe uptake without substantial effects on leukocyte or differential counts (19). If weight is monitored and dactinomycin administration is stopped when weight loss occurs and until weight is regained, the drug can be given for long periods. With this background we started to treat 3-4-month-old B/W female mice with increasing doses to a level that eliminated circulating reticulocytes. Most of the animals in the first series never developed significant proteinuria, and all survived to at least 15 months of age (20) (Figure 1). Control age-matched animals had all died by 12 months. All had heavy proteinuria before they died, and post-mortem examination documented severe renal lesions.

Most of the dactinomycin-treated animals were very long-lived: the median age at death was about 20 months. Many of these mice had tumors, but few had renal lesions more severe than those of female CBA and C57BL/6 mice of similar age (21). Fluorescence

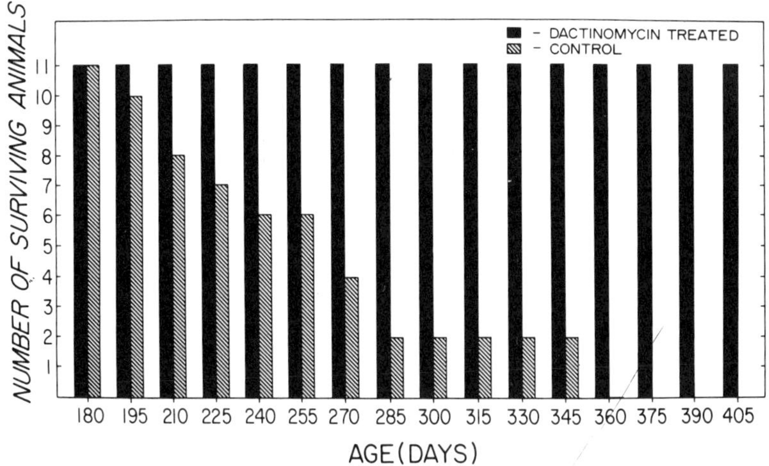

Figure 1. Longevity of dactinomycin-treated and control mice to 15 months. All control animals died by 12 months of age, and all the treated mice survived at 15 months. Reprinted from *Nature* (20) by permission.

microscopy (Figure 2) illustrates the profound involvement of an enlarged glomerulus from an untreated B/W mouse compared to the limited areas of mesangial deposition in both a treated B/W mouse and a C57BL/6 animal of comparable age.

In 1970, Floersheim (15) determined the erythrosuppressive and immunosuppressive capability of a wide range of agents. The dose of each agent was 25% of the LD_{50}. Erythrosuppression was assayed as inhibition of ^{59}Fe uptake, and immunosuppression was assayed as inhibition of Jerne plaque formation; thus each agent was characterized by a single number and by a single point in a scattergram. Dactinomycin, vincristine, vinblastine, and prednisolone were all much more erythrosuppressive than immunosuppressive (Figure 3).

Vincristine has been given for more than a year to a group of B/W mice in the dose of the Floersheim experiment. Hematocrit is monitored closely, and administration is stopped when it drops rapidly in most animals of the group. When hematocrit levels in the range of 40% are regained, treatment is resumed. Most of the animals are alive at 16 months of age or more (Figure 4), in good condition, with insignificant proteinuria. Median survival has been increased by more than 115%, from 238 days to 523 days. In a second group of mice

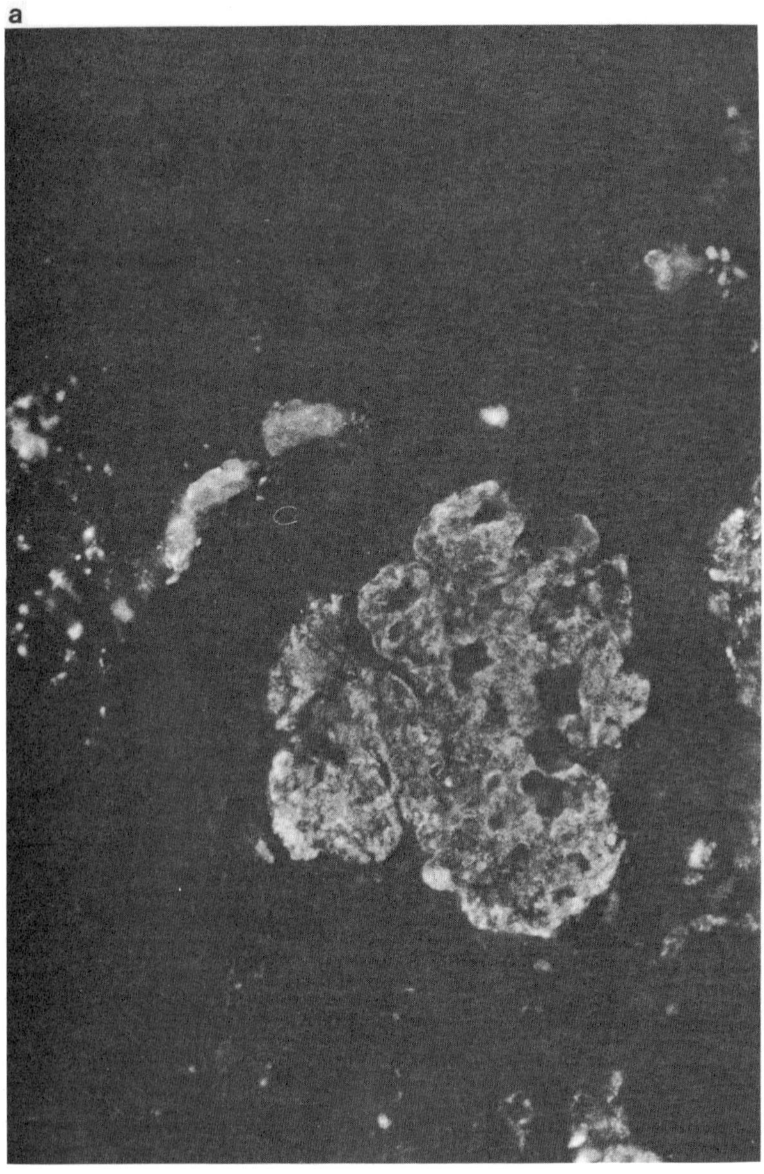

Figure 2. A comparison of fluorescence micrographs of kidney from (a) an untreated B/W mouse with advanced renal disease; (b) a dactinomycin-treated long-lived (573 days) B/W animal; and (c) an aged (17 months) C57BL/6 mouse. (a) shows the extensive involvement of the enlarged glomerulus of the nephritic mouse, and contrasts

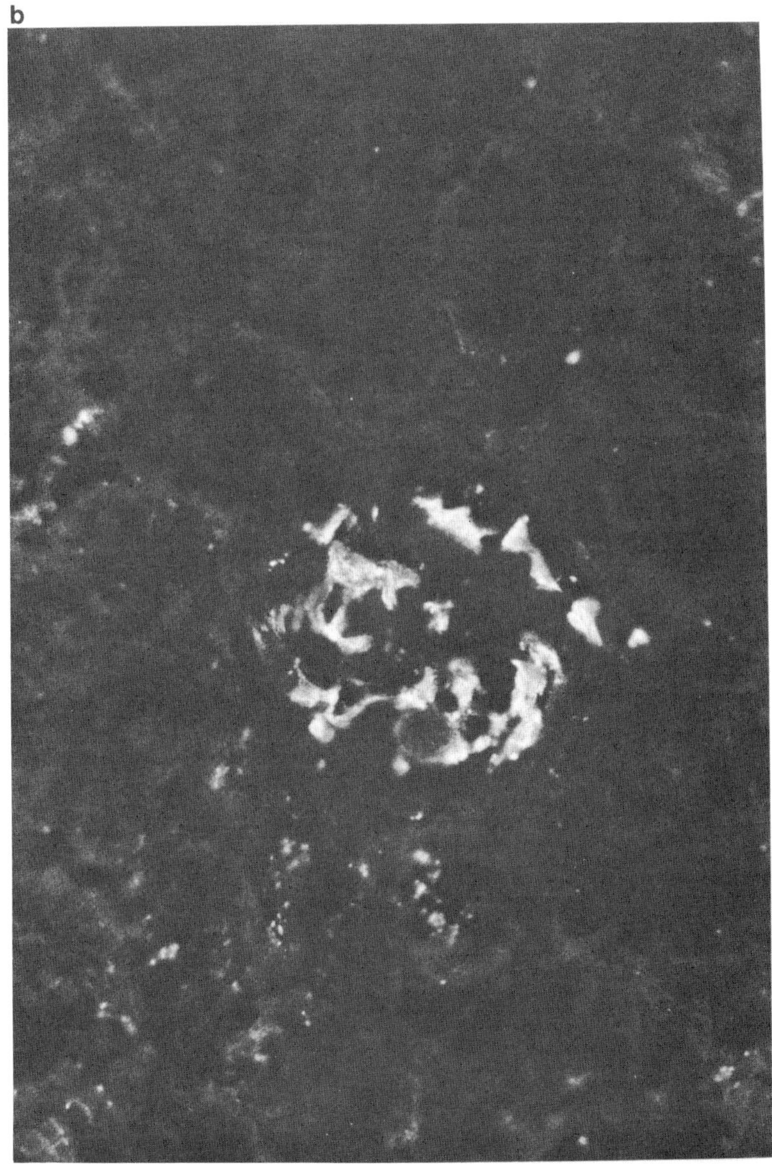

with the similar limited immunoglobulin deposition in the mesangia of both aged mice. The fluorescein-labeled antibody was anti-mouse-IgG in this instance, but the distribution of the other immune reactants was similar.

c

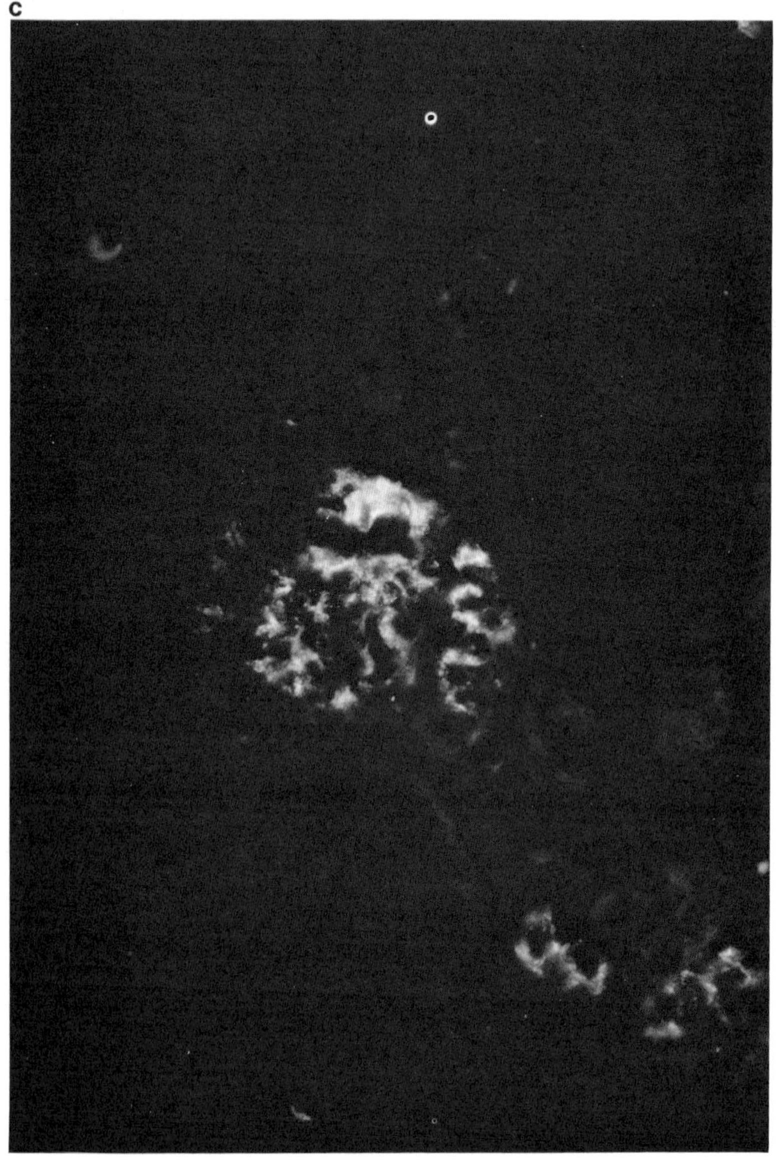

Figure 2c.

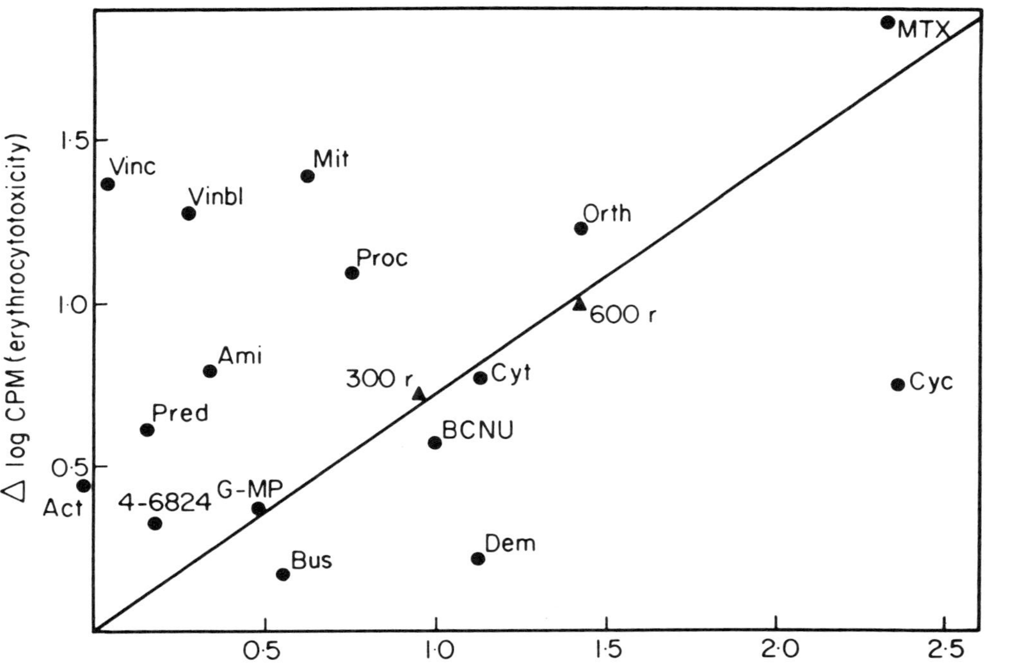

Figure 3. Scatter diagram showing the relative inhibition of erythropoiesis (vertical axis) and antibody response (horizontal axis) by a series of antineoplastic drugs. The diagonal was defined by the two doses of irradiation. It will be seen that dactinomycin (Act), mitomycin (Mit), vincristine (Vinc), vinblastine (Vinbl), and prednisolone (Pred) were more inhibitory of erythropoiesis than of antibody production, and that cyclophosphamide (Cyc) was at the immunosuppressive extreme. Reprinted from *Clinical and Experimental Immunology* (15) by permission.

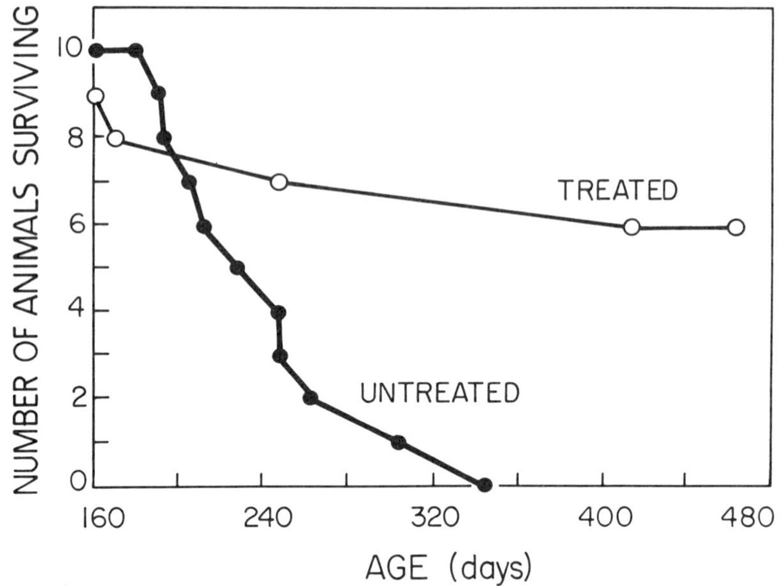

Figure 4. Survival of vincristine-treated B/W mice and aged-matched control animals. Seven of nine mice lived to 400 days or more, and six were living at 500 days.

(Figure 5) we halved the dose of vincristine; survivals were extended by about three months. Median age at death was 304 days in the treated mice and 210 in the controls. Four mice of this group survived at ages of 12 months or more.

The only agent that rivals dactinomycin and vincristine in treatment of B/W disease is cyclophosphamide in bolus doses (22). Cyclophosphamide was, interestingly, the most immunosuppressive agent in the Floersheim series mentioned above, but it also had erythrosuppressive capability at that dose.

In summary, we have proposed that the pathogenesis of B/W disease involves DNA from maturing erythroid cells and have suggested that erythrosuppression may be more effective treatment than immunosuppression. In experiments involving hypertransfusion, dactinomycin treatment, and vincristine administration the data support that hypothesis. If the mechanisms in SLE and B/W disease are similar, our results suggest that there are therapeutic alternatives in SLE that might offer greater benefit with fewer costly side effects.

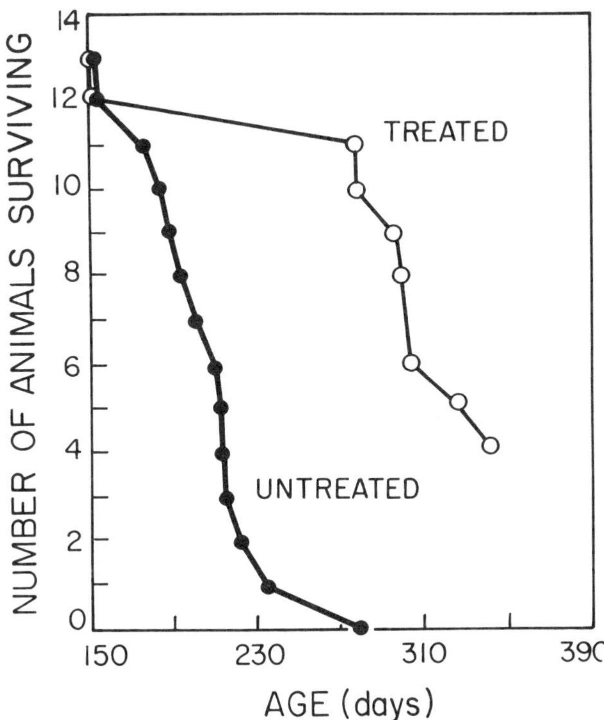

Figure 5. Survival of vincristine-treated B/W mice and age-matched control animals. The dose of the drug was half that of the first experiment, illustrated in Figure 4. Longevity was increased significantly in the treated mice, but much less than in the high-dose group.

ACKNOWLEDGMENTS

Since the experiments described herein span four years, it did not seem feasible to include all of my collaborators as authors. Ms. Carol T. Olsen has participated in all the experiments and her contribution is unique. Mr. Ulrich H. Rudofsky and Dr. Rodrigo E. Urizar, and more recently Mr. Andrew D. Simmons, have helped us with the light or fluorescence microscopic evaluation of the kidneys. I am grateful to all of them, and to Miss June Smith of the University of Minnesota for her help in setting up our mouse colony and training our personnel.

LITERATURE CITED

1. B. J. Helyer and J. B. Howie, Nature 197, 197 (1963)
2. P. H. Lambert and F. J. Dixon, J. Exp. Med. 127, 507 (1968)

3. R. C. Mellors, T. Shirai, T. Aoki, R. J. Huebner and K. Krawczynski, J. Exp. Med. 133, 113 (1971)
4. B. P. Croker, B. C. Del Villano, F. C. Jensen, R. A. Lerner, and F. J. Dixon, J. Exp. Med. 140, 1028 (1974)
5. J. A. Levy, Science 182, 1151 (1973)
6. N. Talal and A. D. Steinberg, Curr. Top. Microbiol. Immunol. 64, 79 (1974)
7. J. A. Levy, Am. J. Clin. Path. 62, 258 (1974)
8. J. East, J. J. Harvey and R. J. Tilly, Clin. Exp. Immunol. 24, 196 (1976)
9. S. K. Datta and R. S. Schwartz, Nature 263, 412 (1976)
10. R. S. Krakauer, T. A. Waldmann, and W. Strober, J. Exp. Med. 144, 662 (1976)
11. J. C. Roder, D. A. Bell, and S. K. Singhal, Cell. Immunol. 29, 272 (1977)
12. S. Kysela and A. D. Steinberg, Clin. Immunol. Immunopathol. 2, 133 (1973)
13. R. E. Wolf and M. Ziff, Arth. Rheum. 19, 1353 (1976)
14. A. E. Gabrielsen, Lancet II, 1116 (1974)
15. G. L. Floersheim, Clin. Exp. Immunol. 6, 861 (1970)
16. T. B. Dunn, R. A. Malmgren, P. G. Carney, and A. W. Green, J. Nat. Cancer Inst. 36, 1003 (1966)
17. P. Tambourin, F. Wendling, N. Barat, and F. Zajdela, Nouv. Rev. Franc. Hemat. 9, 461 (1969)
18. A. E. Gabrielsen and C. T. Olsen, Immunology 33, 449 (1977)
19. K. R. Reissmann and K. Ito, Ann. N.Y. Acad. Sci., 149, 193 (1968)
20. A. E. Gabrielsen, A. S. Lubert and C. T. Olsen, Nature 264, 439 (1976)
21. A. E. Gabrielsen, and U. H. Rudofsky, Fed. Proc. 36, 1296 (1977)
22. P. J. Russell and J. D. Hicks, Lancet I, 440 (1968)

Infection, Immunity, and Genetics
Edited by Herman Friedman, T. Juhani Linna, and James E. Prier

GENETIC CONTROL OF THYROIDITIS

Joseph H. Kite, Jr.

In attempting to understand the cause of autoimmune thyroiditis in humans, a number of experimental animals have been studied. Autoimmune thyroiditis can be produced in a variety of animals, including rabbits, guinea pigs, mice, rats, and monkeys, by the injection of normal thyroid extract plus Freund's complete adjuvant (1). However, one of the more exciting events in experimental studies has been the finding of spontaneously occurring autoimmune disease in animals. The occurrence of autoimmune hemolytic anemia in New Zealand Black (NZB) mice (2, 3) was the first such model that has been explored extensively. A second model that has offered an opportunity for a variety of approaches is the spontaneous development of autoimmune thyroiditis in Obese strain chickens (4, 5, 6). This review is limited to research studies that the authors and others have done with the Obese strain of White Leghorn chickens. Emphasis is placed upon the role of genetics and the immune response. Several reviews can be sought for additional findings (7, 8, 9).

In Obese strain chickens, thyroiditis is so severe that the birds have characteristic phenotypic signs. These were first observed by Dr. R.K. Cole, a geneticist at Cornell University (10,11). The signs are a smaller skeletal structure, long silky feathers, a smaller comb and wattle, extra subcutaneous fat, and poor laying ability (Figure 1). All of these characteristics reflect a hypothyroid condition. For ease of communication, the authors (5) designated the strain "Obese strain" (OS), although this term reflects only one characteristic of the hypothyroid condition. In Figure 2, the Obese strain chicken on the right is compared with a normal male, each about 5 months of age. The reduced size of the comb and wattle is quite obvious in the OS chicken. When Dr. Cole first recognized the several phenotypic characteristics in less than 1% of chickens in a normal closed flock of White Leghorn chickens, he attempted to select for these charac-

---

Supported by USPHS NIH Grant CA-02357 and 5-S07-RR7066, BRSG from State University of New York at Buffalo.

Figure 1. Obese strain chicken, female, 5 months of age.

Figure 2. Obese strain chicken, male, 5 months of age (right), and C strain chicken, male, 5 months of age (left).

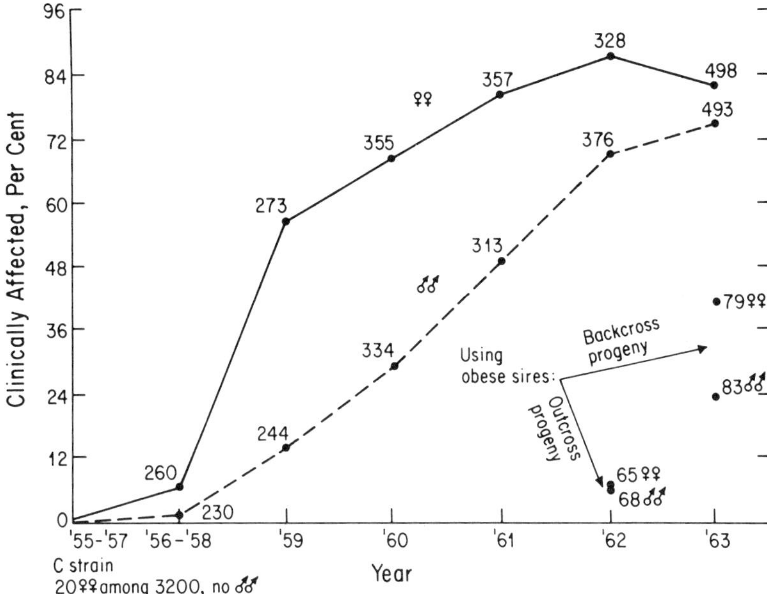

Figure 3. The effect of selection on the frequency of obesity within the C strain of White Leghorns. The difference between the sexes has become less after several generations of selection. The numeral indicates the number of individuals classified to determine the specific point on the graph. (From Cole, R. K. "Hereditary hypothyroidism in the domestic fowl." *Genetics 53*:1021-1033, 1966).

teristics to produce a strain. Only a few chickens with these phenotypic characteristics were recognized in 1955-56 (Figure 3). At first these hypothyroid chickens were mated to normal chickens of the same flock and later, as males were produced with hypothyroid characteristics, they were used for future breedings. Thus, by selective matings between chickens expressing the phenotypic hypothyroid signs, a strain of chickens was developed with increasing incidence and severity of thyroiditis. At present more than 90% of the chickens of this strain develop hypothyroid characteristics. The trait appears to be polygenic in character since outcrosses result in a rapid loss of the phenotypic signs.

Examination of the thyroid glands of OS chickens for pathology by light microscopy revealed that initially there may be a small focus of infiltration by lymphocytes and mononuclear cells. This finding is rated a 1+ infiltration, with less than 25% of the cross-sectioned area of the thyroid gland involved (Figure 4). Figure 5 demonstrates a

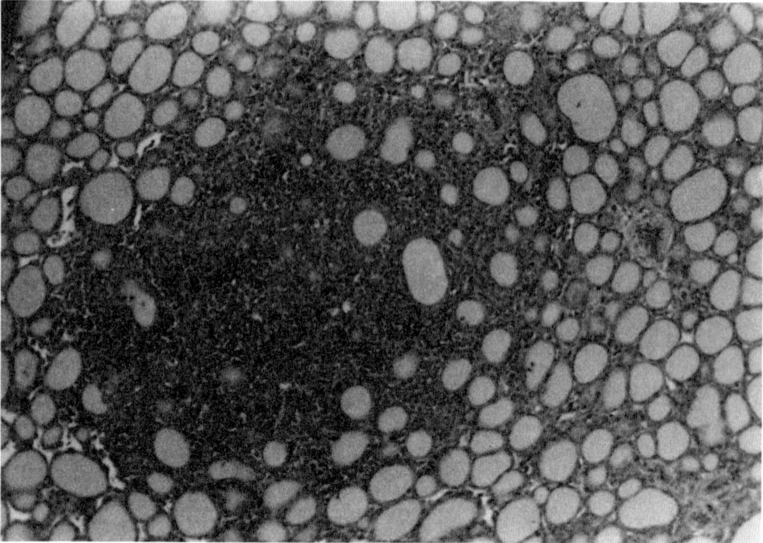

Figure 4. Histological appearance of thyroid glands from Obese strain chickens: 1 + infiltration.

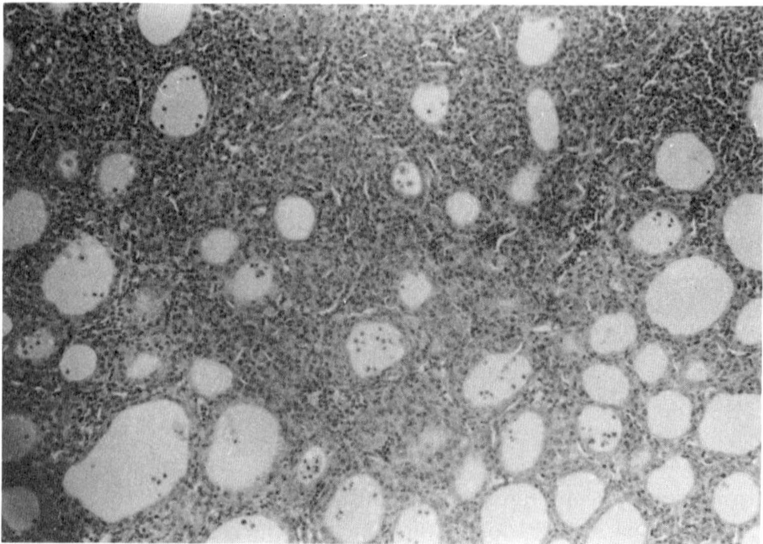

Figure 5. Histological appearance of thyroid glands from Obese strain chickens: 2 + infiltration.

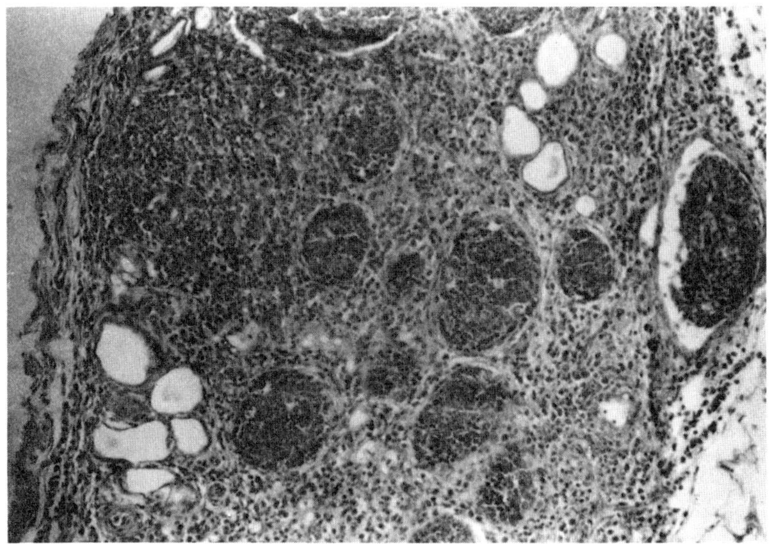

Figure 6. Histological appearance of thyroid glands from Obese strain chickens: 4 + infiltration.

later stage designated 2+, invasion of 50–75% of the area. Figure 6 shows a 4+ infiltration, almost 100% of the thyroid involved, with destruction of the thyroid gland so that very few follicles remain. Many germinal centers are found containing lymphocytes, plasma cells, and macrophages.

The time sequence for the development of thyroiditis is illustrated in Table 1. At the time of hatching there is no histological change in the thyroid, as observed by light microscopy. After 1 week following hatching, there is occasionally a small focus of infiltration in the thyroid. This will progress so that between 3 and 6 weeks one may observe a maximum degree of thyroiditis developing. The C strain chicken, the parent strain from which the OS was derived, has little or no histological change in the thyroid glands. The hemagglutination titer is uniformly negative. Serum antibodies to thyroglobulin can be detected in OS chickens by passive hemagglutination. As individual chickens are bled periodically, a rise in titer is frequently observed in the serum (5).

Since antibodies to thyroglobulin were detected in the sera of OS chickens on the day of hatching, antibodies were then sought in various components of embryonating eggs. Examination of the serum

Table 1. Temporal development of thyroiditis

Time of sacrifice	Obese strain chickens								C strain chickens		
	Histology of thyroid				Hemagglutination titer				Histology of thyroid		Hemaggluti-nation titer
14-day embryo	–[a]	–	–						–	–	<2
1 day	–	–	–	–	40[a]	80	20	80	–	–	<2
1 week	–	–	+	–	<2	20	20	20	–	–	<2
2 weeks	+ +	+ + +	+	+	<2	10	5	<2	+	–	<2
3 weeks	+ +	+ +	+ + + +	+ + +	<2	<2	<2	<2	+	–	<2
4 weeks	+ +	+ +	+ + +	–	80	20	160	2	+	+	<2
7 weeks	+ + + +	+ + + +	+	+ + +	40	40	2	160	+	–	<2
9 weeks	+ + + +	+	+ + + +	+ + + +	<2	<2	<2	<2	–	+ +	<2
12 weeks	+ + +	+ + +			2	5			–	+	<2
16 weeks	+ + + +	+ + + +	+ + +	+ + +	<2	5	<2	40	+	+	<2
20 weeks	+ + +		+ + +	+ + +	<2	<2	<2	<2	–	–	<2
24 weeks	+ + +	+ + +	+ + +		320	<2	5		+	+ +	<2

[a] *Each* + or – symbol represents results from birds killed at the time interval indicated.

at 12 to 19 days revealed the presence of antibodies while the yolk contained antibodies at all time intervals tested (Table 2). Furthermore, examination of the parent hen revealed frequently a high titer in the serum and in the follicles, the latter destined to become the yolk of the developing eggs. Thus passive transfer of thyroglobulin antibodies could be demonstrated from the parent hen to the developing embryo through the yolk. It was of concern whether these antibodies could be cytotoxic and thus responsible for an initial injury to the embryo during the 21 days of egg incubation. Passive transfer of high titered OS chicken serum to normal embryos, to normal chicks, or to normal thyroid cells cultured in vitro has not led to significant thyroid pathology or cell damage. Furthermore, eggs laid by some hens that do not have detectable levels of serum antibodies may still result in chicks that develop severe thyroiditis. At first these findings could suggest that thyroglobulin antibodies are the result of thyroid injury and not the cause. However, further studies on the role of antibody alone, in combination with null cells (K cell cytotoxicity), or its participation in immune complexes must be studied to evaluate accurately its role in this autoimmune disease.

Antibodies to thyroglobulin can be detected also by precipitation in agar gel. Furthermore, these serum antibodies are autoimmune in nature because they react with an autologous thyroid extract (5). Antibodies can be found in serum to other tissue organs in addition to thyroid (Figure 7). Autoantibodies to liver and kidney are occasionally found in older OS chickens, but rarely antibodies to other organs. No significant damage to these organs has been observed histologically. However, these findings do demonstrate that the autoimmune reactivity in OS chickens is not limited solely to the thyroid. Thus in attempting to understand the mechanism(s) responsible for thyroiditis, one must consider that the immune response is not restricted to thyroid antigens.

Table 2. Antibodies to thyroglobulin in embryonating eggs from OS chickens

Materials tested	Day of incubation	Number tested	Number positive	Highest titer
Serum	7- 9	7	0	<2
	12–19	28	15	128
Yolk	0–19	45	30	2560
Allantoic fluid	7–14	9	0	<2
Amnionic fluid	7–19	11	0	<2

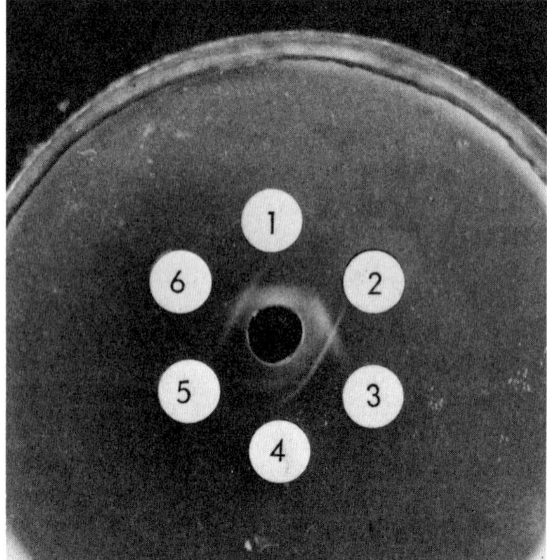

Figure 7. Gel precipitation test. Autoantibodies are demonstrated by the reaction of Obese strain serum with the chicken's own tissue extracts. Center well contains Obese strain serum. Outer wells contain autologous extracts of: 1—thyroid (undiluted), 2—thyroid (1:10), 3—liver (undiluted), 4—liver (1:10), 5—kidney (1:10), 6—kidney (undiluted).

The chicken offers an invaluable tool for the immunologist since the bursa of Fabricius and the thymus are clearly separated organs which are responsible for the maturation of T-dependent and B-dependent lymphoid cells. Surgical bursectomy at the time of hatching greatly reduces the incidence of thyroiditis (12; Table 3). The administration of an androgen such as testosterone propionate (TP) or cyclophosphamide (CY) will reduce further the severity of thyroiditis. Thus, spontaneous autoimmune thyroiditis is dependent upon bursa-derived cells. These findings suggest that new attention be given to the role of thyroglobulin antibodies, acting in concert with so-called null or K cells or by localization in immune complexes.

From evidence of experimental thyroiditis in other animals, it was expected that thymectomy might lead to a reduction in the severity of thyroiditis. However, surgical thymectomy performed on the day of hatching increases the severity of thyroiditis.(13). This evidence would favor a role for suppressor T cells in the containment of thyroiditis, although the failure to obtain severe thyroiditis in normal chickens by thymectomy would suggest that other factors are also essential.

Table 3. Bursectomy and thymectomy of OS chickens

Treatment	Thyroid pathology
Untreated	2-3 +
Bursectomy	
Hormonal (14-day embryo)	None
Surgical (day of hatching)	1 +
Thymectomy	
Surgical (day of hatching)	4 +

The prediction of genotypes was made from transplantation studies in offspring within matings of each strain, OS and CS (14). In both strains two or three males were each mated to several females. From each mating chicks were tested for skin grafting and graft versus host reactions. The chickens were judged homozygous or heterozygous for two alleles based on their acceptance of grafts and tests for splenomegaly. Arbitrary designations of the major genotypes were made: B^1, B^2, B^3, B^4.

Antisera specific for various B alleles were produced by injection of erythrocytes from chickens of one genotype into another, designated as donor and recipient (Table 4). After appropriate absorption, the antisera were then specific for various genotypes. Using this specific antisera, the frequency of different B alleles in OS birds was determined (Table 5). In the OS, two B alleles, B^1 and B^4, were identified with approximately equal gene frequency. B^3 was observed only rarely. Only one of the frequent alleles in OS chickens, B^1, was identified in the parent CS. The B^3 allele and another allele, B^2, were present in moderate frequency in CS chickens. Matings of B^1B^4 OS heterozygous birds resulted in the expected genotype ration of 1:2:1 (one B^1B^1: two B^1B^4: one B^4B^4).

The next study was conducted to determine whether the B genotype influences the susceptibility of OS chickens with spontaneous autoimmune thyroiditis. OS birds of the genotypes B^1B^1, B^1B^4, and B^4B^4 were killed at 3, 6, and 10 weeks of age and the mean thyroid pathology was evaluated on the scale of +1 to +4 (Figure 8). Findings showed that greater pathology occurred in B^1B^1 and B^1B^4 birds than in their B^4B^4 siblings. Antibodies to thyroglobulin were also measured in the sera of these birds and were found higher than in the genotypes B^1B^1 and B^1B^4. Thus these results indicated that the alleles in the B locus, or genes closely linked to it, influence the immune response in spontaneous autoimmune thyroiditis.

Table 4. Specificity of antisera to the B antigens in OS and CS chickens

Donor[a]	Recipient	Antiserum	Unabsorbed antiserum reacts with	Antiserum absorbed with	Reagent specific for
CS B^1B^2	CS B^2B^2	B1	B_1, B_3, B_4, B_5,	B_3, B_4	B_1, B_5
CS B^2B^3	CS B^1B^3	B2	B_2	None	B_2
CS B^2B^3	CS B^1B^2	B3	B_3, B_4,	B_4	B_3
OS B^4B^4	OS B^1B^1	B4	B_4, B_3	B_3	B_4

From Bacon, L. D., Kite, J. H., and Rose, N. R. 1973. Immunogenetic detection of B-locus genotypes in chickens with autoimmune thyroiditis. *Transplantation 16*:591–598.

[a] Erythrocytes from donors used for injection of recipients.

Table 5. Genotypes of OS and CS chickens

Genotype	Frequency	
	OS	CS
B^1	0.50	0.51
B^2		0.19
B^3	0.04	0.26
B^4	0.46	
B^5		0.04

In these studies continued by Bacon and Rose (9), thyroiditis was measured in F_1 chickens produced by mating OS and CS chickens (Table 6). Again it appeared that the genotype of the F_1 chicks determined the severity of the disease. The majority of the birds with a B^3 allele developed significant disease. The strain derivation of the B allele, whether OS or CS, did not influence the results. It is con-

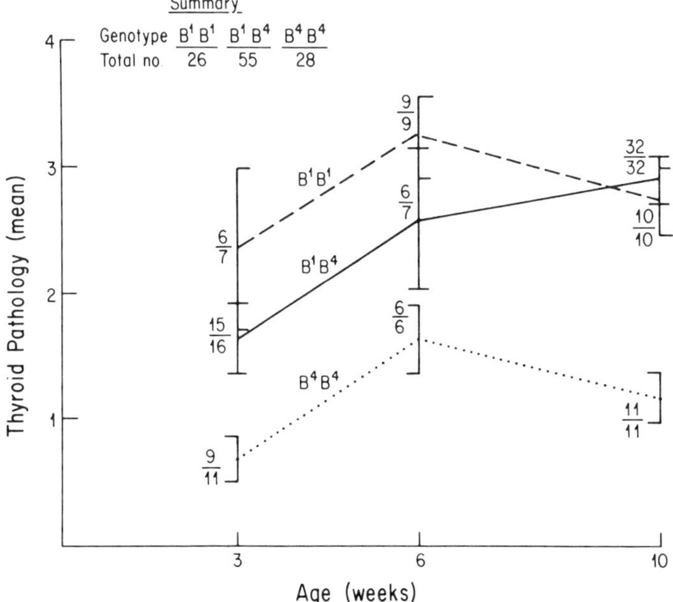

Figure 8. Thyroid pathology of chicks from $B^1B^4 \times B^1B^4$ OS matings. The fraction at each point indicates the ratio of number of birds with lymphocyte infiltration of the thyroids to the total number of birds. (From Bacon, L. D., Kite, J. H., and Rose, N. R. "Relation between the major histocompatibility (B) locus and autoimmune thyroiditis in obese chickens." *Science 186*:274–275. 1974).

Table 6. Thyroiditis in (OS × CS) F_1 chickens

B genotype	Number of birds	Pathology $\bar{x} \pm$ SE	Antibody $\bar{x} \pm$ SE
B^3B^3	11	2.0 ± 0.5[a]	5.5 ± 1.4[b]
B^1B^3	33	1.4 ± 0.2	4.6 ± 0.7
B^2B^3	17	1.6 ± 0.3	6.6 ± 1.1
B^3B^4	22	1.9 ± 0.3	5.5 ± 1.0
B^1B^1	18	0.5 ± 0.1	2.5 ± 0.7
B^1B^4	25	1.0 ± 0.2	2.8 ± 0.7
B^1B^2	9	0.2 ± 0.0	0.8 ± 0.2
B^2B^4	16	0.2 ± 0.0	0.6 ± 0.3

From Rose, N. R., Bacon, L. D., and Sundick, R. S. 1976. Genetic determinants of thyroiditis in the OS chicken. *Transplant. Rev. 31*:264–284.
[a] *Mean* ± SE of thyroid pathology. 0.2 unit = 5% infiltration.
[b] Mean ± SE of $\log_2$ passive hemagglutin in titer to thyroglobulin.

cluded that in crosses from a susceptible and a normal animal (C strain) the major histocompatibility complex is a most important factor in determining the severity of autoimmune disease.

It was recognized that there might be differences in the *B* alleles that were not detected serologically. Therefore, OS and CS chickens were compared in graft versus host reactions and by mixed lymphocyte reactions. Using both of these techniques, no significant reaction could be detected between B^1B^1 of OS and B^1B^1 of CS, strongly suggesting identity of the two strains with regard to the B^1 alleles (9).

There are two additional points that should be emphasized. As indicated previously, a 6-week-old B^1B^1 chicken has severe disease whereas F_1 B^1B^1 chickens have little thyroiditis. However, homozygous matings of OS B^4B^4 have shown greater thyroiditis at 10 weeks of age than B^4B^4 birds that are from heterozygous matings. Thus these points would suggest that additional genes modify spontaneous autoimmune thyroiditis (9).

The structure of the B histocompatibility complex has not been characterized in chickens, although genes controlling the immune response to several antigens have been linked to the *B* locus. Assuming that *Ir* genes are responsible for susceptibility to autoimmune thyroiditis, Rose, Bacon, and Sundick (9) propose that: 1) B^3 alleles determine a high response to antigenic determinants of thyroglobulin; 2) B^1 alleles determine moderate response to antigenic determinants of thyroglobulin; and 3) B^2 and B^4 alleles determine low or no response to antigenic determinants of thyroglobulin. It can be conceived that non-responder birds could recognize antigenic determinants but preferentially develop suppressive function rather than immune response. The most likely candidate for the suppressive gene in OS chickens is the B^4 allele or one closely linked to it.

Although we do not know the fine structure of the major histocompatibility locus in the chicken, if the alleles just described were the only important factors, then we might expect to achieve a successful transfer of thyroiditis by injecting viable spleen cells from an OS B^1B^1 donor into a CS B^1B^1 recipient. In Table 7 it is seen that this type of transfer has not been successful. In only 3% (12/361) of transfers of spleen cells from OS chickens to CS chickens of the same genotype (B^1B^1) did significant thyroiditis occur. The mean pathology of the 12 chickens that did develop thyroiditis was 2.3, a degree of infiltration considered minimal.

In our experience, the only successful means of transferring thyroiditis has been by injecting spleen cells from an OS B^1B^1 donor into an OS bursectomized chicken of the B^1B^1 genotype. Thus the donor and recipient are of the same major genotype, but because of bursectomy the recipient does not develop thyroiditis spontaneously. In this type of transfer 44%, or 61 of 141 transfers, were successful. Thyroids from these successful transfers had a mean pathology of 2.4. The failure to transfer thyroiditis from OS to CS chickens would suggest that more factors are needed to establish an autoimmune disease than the singular presence of lymphoid cells capable of producing antibody to thyroglobulin or capable of reacting with thyroglobulin.

Consideration is now given to the role of the thymus in regulation of thyroiditis. Earlier we demonstrated that thymectomy of OS chickens leads to increased severity of thyroiditis (13). The question

Table 7. Transfer of spleen cells

Donor	Recipient	
OS (B^1B^1)	CS (B^1B^1)	
Pathology	Transfer positive	Pathology
3.1	3% (12 of 361)	2.3
OS (B^1B^1)	OS − Bx (B^1B^1)	
Pathology	Transfer positive	Pathology
3.3	44% (61 of 141)	2.4

may be asked whether there is a difference in the thymus of OS and CS chickens. Figure 9 summarized an experiment in which skin allografts are exchanged between OS and CS chickens, both of the same B^1B^1 genotype (9). The mean survival times of the grafts were similar in OS and CS chickens, reflecting the belief that these alleles are the same in both strains. If neonatal thymectomy is done on CS birds, then the mean survival time of skin allografts is increased as expected. In contrast, thymectomy of OS birds has very little effect. Only one of 16 skin grafts survived 15 days after transplantation. One reason that has been postulated for the lack of effect following the thymectomy of OS birds is that the effector cells have already left the thymus and are located in peripheral lymphoid tissues.

Our current studies in this area support the view that OS chickens have a hyperreactivity to homologous and heterologous antigens. It appears that the B cells have become activated earlier than in CS or normal chickens, perhaps because of an earlier maturation or stimulation of the cells. Three approaches have been taken to ex-

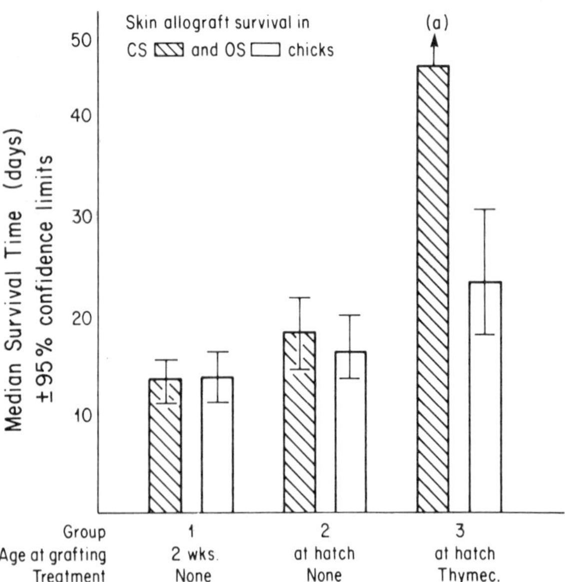

Figure 9. Intrastrain skin grafts between B^1B^1 OS and B^1B^1 CS chickens. More than half of the grafts (designated a) survived 50 days and therefore the median survival time could not be calculated by the Litchfield method. Data of Jakobisiak, Sundick, Bacon, and Rose. (From Rose, N. R., Bacon, L. D., and Sundick, R. S. "Genetic determinants of thyroiditis in the OS chicken." *Transplant. Rev.*, *31*:264-284, 1976).

amine the cellular response of OS chickens. The first two procedures have been performed by Mr. John Tyler and the third in collaboration with Dr. Jane Pascale. The first approach was to measure lymphocyte cell stimulation by mitogens. In this experiment (Table 8), OS, CS, and two normal inbred lines of chickens were studied by comparing the response of peripheral blood leukocytes (PBL), spleen, and thymus cells to phytohemagglutinin (PHA) and Concanavalin A. The response of PBL from OS was always higher, so instead of listing counts per minute, OS PBL were assigned the maximum response of 100% and all the other counts were compared to it. If we examine the mean values for all three cell populations, it is seen that OS is the maximum responder, CS is the next most reactive, and the two normal lines are considerably less.

A second method of examining cellular response is to look for antibodies to foreign antigens, i.e., sheep red blood cells. In Table 9, a priming dose was given on days 4 and 11 and a challenged dose on day 22. The response of CS is a dose-independent response with a gradual increase in titer especially observed in concentrations of 10^8

Table 8. Cellular stimulation by mitogens

		PHA (% max)	Con A (% max)
OS	PBL	100	100
	Spleen	86	95
	Thymus	46	100
		$\bar{x}$ = 77	$\bar{x}$ = 98
CS	PBL	64	90
	Spleen	100	100
	Thymus	29	38
		$\bar{x}$ = 64	$\bar{x}$ = 76
KSU	PBL	40	16
	Spleen	18	51
	Thymus	27	40
		$\bar{x}$ = 28	$\bar{x}$ = 36
Line G	PBL	7	3
	Spleen	78	15
	Thymus	100	24
		$\bar{x}$ = 62	$\bar{x}$ = 14

Table 9. Antibody response to sheep red blood cells[a]

SRBC	Strain	Log2 titer Age in days 11	14	18	21	25	27
10^8	OS	1.8	5.7	4.7	4.5	6.5	8.5
	CS	0.6	1.6	1.6	2.0	2.8	5.8
10^9	OS	2.7	6.2	5.0	5.4	6.6	8.8
	CS	2.5	3.8	2.8	2.4	2.6	4.2
2×10^9	OS	2.2	5.8	4.2	4.8	7.5	8.7
	CS	3.2	5.7	4.0	4.2	5.2	6.2

[a] Priming doses: days 4 and 11. Challenge dose: all groups received 10^8 cells on day 22.

and 10^9 sheep red blood cells. However, OS chickens showed a higher antibody response early.

Finally, a third procedure involved the response of chickens to growth of Rous sarcoma (Table 10). When Rous sarcoma virus (RSV) was injected 5 days after hatching, no significant reduction in tumor growth occurred in OS or CS or in normal strains of chickens. However, when RSV was injected at 2 ½ weeks or 6 weeks following hatching and corresponded to the maximum time of autoimmune response, then there was considerable decrease in the growth of tumors.

An analysis of these three approaches is complicated by the fact that an autoimmune thyroiditis is occurring simultaneously with measurements of immune response to other antigens. When each of these experiments was repeated in OS chickens thyroidectomized by

Table 10. Growth in chickens of tumors induced by Rous sarcoma virus: variation in time of RSV injection

Chicken strain	Time of RSV[a] injection after hatching: 5 days	2.5 weeks	6 weeks
Obese strain	10.3[b]	2.1	1.5
C strain	15.2	48.0	19.4
Line G	25.3	41.8	20.1

[a] RSV, lot 1, 10^{-3} dilution.
[b] Size of tumor produced in wing web, measured in cc, 3 weeks after RSV injection. This figure is the mean value from five chickens.

fulguration on the day of hatching, hyperreactivity was still observed in the absence of a thyroid autoimmune response. These findings would suggest that a hyperimmune response in OS chickens is a general phenomena, not selective only to thyroglobulin, and may be genetically determined.

Another possible cause of spontaneous autoimmune thyroiditis could be a defect in the thyroid glands which may or may not involve virus infection. Evidence to support this view has been cited by: 1) degenerative changes in the thyroid of OS chickens detected by electron microscopy as early as one week of age (15), 2) presence of particles resembling virus in thyroids from OS chickens (15), and 3) endocrine abnormalities of the thyroid gland. Such abnormalities have usually been detected some time after hatching, so that they may reflect the consequence of thyroid destruction. We demonstrated that the T_3, and especially the T_4 component, thyroxine, are almost absent from the serum of OS chickens (16). A summary of studies on the detection of thyroid hormones in the serum of OS birds has been performed in collaboration with Drs. B.N. Premachandra and S. Lang at the Veteran's Administration Hospital, St. Louis, Missouri (Table 11). In the first group, Obese strain chickens that have been untreated or treated by surgical thymectomy have diminished T_3, T_4, and free T_4 hormones in the serum. In group two, normal chickens represented by inbred strain KSU, CS, the parent strain of the OS, or normal random bred chickens presented normal ranges of hormones in the serum. Obese strain that has been bursectomized by TP or CY treatment and had no thyroiditis also demonstrated normal ranges of these hormones. Rose, Bacon, and Sundick (9) have shown that OS chickens had the highest uptake of ^{131}I, three times that of normal strains or twice the uptake per mg of thyroid of normal strains, and yet they had the lowest concentration of thyroxine.

Table 11. Thyroid hormones in the sera of OS chickens

Group	Strain	Treatment	T_3	T_4	Free T_4
I	OS	None	107	0.34	0.09
	OS	Tx	107	0.34	0.02
II	KSU	None	211	0.75	0.21
	CS	None	207	0.68	0.15
	NRB	None	255	0.90	
	OS	Bx-TP	161	0.93	0.29
	OS	Bx-CY	195	0.69	0.19

In attempting to correlate these findings and additional ones of other investigators, insufficient data are available at the moment to prove any one of several possible hypotheses. However, our findings have narrowed the possibilities to some degree. Since bursectomy prevents spontaneous autoimmune thyroiditis, attention is directed to the humoral (B cell mediated) aspects of the cellular immune reponse rather than to the direct effects of T cell-mediated immunity. Antibody in conjunction with complement, immune complexes, or antibody plus the action of null cells (K cell cytotoxicity) are strong candidates for initiation of cell injury, although direct B cell-mediated injury must be given consideration. Thymectomy of OS chicks leads to an increased severity of thyroiditis, suggesting that suppressor T cells have a controlling role in the final expression of the disease. Although the blood group B major histocompatibility antigen is not well defined, there is strong evidence for an association of certain alleles with the severity of thyroiditis. One may hypothesize several ways in which immune response genes may affect the expression of thyroiditis: 1) Alteration of thymus cell membranes could expose new histocompatibility antigens or viral receptors that would allow infection by a ubiquitous virus. Unaltered cell populations of T cells may recognize these new antigens and be stimulated to respond. If these stimulated populations contain helper T cells, they may be able to react with B cells binding autoantigen. 2) Modification of the bursa cell membranes may expose viral receptors and permit infection of the bursa by a ubiquitous virus, leading to a nonspecific stimulation of B cells that would bypass a requirement for helper T cells. It is known that Epstein-Barr virus can stimulate B cells in this manner. 3) Modification of thymus cell membranes with or without concomitant viral infection could lead to autoantibody production to thymus and destruction of suppressor T cells. 4) Modification of the thyroid cell membranes, with or without viral infection, could also trigger an autoimmune response.

In the final analysis, it appears that several general phenomena may contribute to the expression of spontaneous autoimmune thyroiditis. Genetic and immunologic factors are clearly important, whereas the role of a viral infection remains a possible but unproven contributing factor. The loss of suppressor T cells does seem to influence the overall expression of thyroiditis in OS chickens but may not alone be the principal causative factor. A primary defect in the thyroid gland itself would not account for other autoimmune responses. The primary initiating factor in the disease may be understood following knowledge of how B cells are stimulated and why an autoimmune response develops so strongly to thyroglobulin. An understanding of the final destruction of thyroid parenchymal cells by stimulated B cells should occur following an increased knowledge in this model of the role of antibody to thyroglobulin, participation of immune complexes, and the action of null cells with antibody in the cytolytic process.

LITERATURE CITED

1. Bigazzi, P. E., and Rose, N. R. 1975. Spontaneous autoimmune thyroiditis in animals as a model of human disease. Prog. Allergy 19:245–274.
2. Bielschowsky, M., Helyer, B. J., and Howie, J. B. 1959. Spontaneous haemolytic anaemia in mice of the NZB/B1 strain. Proc. Univ. Otago Med. School 37:9.
3. Howie, J. B., and Simpson, L. O. 1976. The immunopathology of the NZB mice and their hybrids. In P. A. Miescher and H. J. Müller-Eberhard (eds.), Textbook of Immunopathology, pp. 247–278.
4. Cole, R. K., Kite, J. H., and Witebskey, E. 1968. Hereditary autoimmune thyroiditis in the fowl. Science 160:1357–1358.
5. Witebsky, E., Kite, J. H., Wick, G., and Cole, R. K. 1969. Spontaneous thyroiditis in the obese strain of chickens. I. Demonstration of circulating autoantibodies. J. Immunol. 103:708–715.
6. Kite, J. H., Wick, G., Twarog, B., and Witebsky, E. 1969. Spontaneous thyroiditis in the obese strain of chickens. II. Investigations on the development of the disease. J. Immunol. 103:1331–1341.
7. Cole, R. K., Kite, J. H., Wick, G., and Witebsky, E. 1970. Inherited autoimmune thyroiditis in the fowl. Poultry Sci. 49:839–848.
8. Wick, G., Sundick, R. S., and Albini, B. 1974. A review. The obese strain (OS) of chickens: An animal model with spontaneous autoimmune thyroiditis. Clin. Immunol. Immunopath. 3:272–300.
9. Rose, N. R., Bacon, L. D., and Sundick, R. S. 1976. Genetic determinants of thyroiditis in the OS chicken. Transplant. Rev. 31:264–284.
10. van Tienhoven, A., and Cole, R. K. 1962. Endocrine disturbances in obese chickens. Anat. Rec. 142:111–118.
11. Cole, R. K. 1966. Hereditary hypothyroidism in the domestic fowl. Genetics 53:1021–1033.
12. Wick, G., Kite, J. H., Cole, R. K., and Witebsky, E. 1970. Spontaneous thyroiditis in the obese strain of chickens. III. The effect of bursectomy on the development of the disease. J. Immunol. 104:45–53.
13. Wick, G., Kite, J. H., and Witebsky, E. 1970. Spontaneous thyroiditis in the obese strain of chickens. IV. The effect of thymectomy and thymobursectomy on the development of the disease. J. Immunol. 104:54–62.
14. Bacon, L. D., Kite, J. H., and Rose, N. R. 1973. Immunogenetic detection of B-locus genotypes in chickens with autoimmune thyroiditis. Transplantation 16:591–598.

15. Wick, G., and Graf, J. 1972. Electron microscopy studies in chickens of the obese strain with spontaneous hereditary autoimmune thyroiditis. Lab. Invest. 27:400-411.
16. Witebsky, E. 1968. The clinical pathology of autoimmunization. Annual Meeting of the American Society of Clinical Pathologists, Chicago, Illinois. Amer. J. Clin. Pathol. 49:301-311.

Infection, Immunity, and Genetics
Edited by Herman Friedman, T. Juhani Linna, and James E. Prier

Index